T0143154

THE LIFE AND SCIENCE OF HAROLD C. UREY

synthesis

A series in the history of chemistry, broadly construed, edited by Carin Berkowitz, Angela N. H. Creager, John E. Lesch, Lawrence M. Principe, Alan Rocke, and E. C. Spary, in partnership with the Science History Institute

The Life and Science of
HAROLD C. UREY

MATTHEW SHINDELL

The University of Chicago Press • Chicago and London

The University of Chicago Press, Chicago 60637
The University of Chicago Press, Ltd., London
© 2019 by The Smithsonian Institution and Matthew Shindell
All rights reserved. No part of this book may be used or reproduced in any manner whatsoever without written permission, except in the case of brief quotations in critical articles and reviews. For more information, contact the University of Chicago Press, 1427 E. 60th St., Chicago, IL 60637.
Published 2019
Printed in the United States of America

28 27 26 25 24 23 22 21 20 19   1 2 3 4 5

ISBN-13: 978-0-226-66208-4 (cloth)
ISBN-13: 978-0-226-66211-4 (e-book)
DOI: https://doi.org/10.7208/chicago/9780226662114.001.0001

Library of Congress Cataloging-in-Publication Data

Names: Shindell, Matthew, author.
Title: The life and science of Harold C. Urey / Matthew Shindell.
Description: Chicago : University of Chicago Press, 2019. |
Series: Synthesis | Includes bibliographical references and index.
Identifiers: LCCN 2019025459 | ISBN 9780226662084 (cloth) |
ISBN 9780226662114 (ebook)
Subjects: LCSH: Urey, Harold Clayton, 1893–1981. | Chemists—
United States—Biography.
Classification: LCC QD22.U74 S55 2019 | DDC 540.92 [B]—dc23
LC record available at https://lccn.loc.gov/2019025459

♾ This paper meets the requirements of ANSI/NISO Z39.48-1992 (Permanence of Paper).

Contents

INTRODUCTION

The Making and Remaking of an American Chemist

On Thanksgiving Day 1931, Frieda Daum Urey sat down with her two young daughters and a few invited guests to a holiday dinner in the Manhattan apartment she rented with her husband, Harold. The meal would have to start without him. He had been with them earlier in the day, when she and the girls went to the Macy's Thanksgiving Day Parade. Harold was an associate professor of chemistry at Columbia University, and his lab was only a few blocks from their home on Claremont Avenue. On their way to the parade route, he had stopped to check on an experiment. As was often the case, he could not take his mind off his scientific work. The marching bands and outlandish parade balloons did not interest him. Frieda, who had trained as a bacteriologist, understood the scientific life and its demands from firsthand experience. In five years of marriage she had witnessed his uncanny ability to lose track of time when he concentrated on a problem. It was as though he could shut out the world entirely. Now, well past dinnertime, Frieda knew that she had once again lost him to his laboratory. But this evening, which began with such a familiar disappointment, ended in triumph. When Harold did eventually come home, he exclaimed as he entered, "Frieda, we have arrived!"[1]

Indeed, they had arrived. With the help of his collaborators, Ferdinand G. Brickwedde and George M. Murphy, Harold C. Urey had experimentally proved the existence of an isotope of hydrogen with mass 2. It was an isotope that until then had been considered either unlikely to exist or too rare to detect. He had succeeded in producing a concentrated sample and detecting the spectral signature of the elusive heavy isotope of hydrogen.[2] He had solved a technically sweet problem, and it came with great rewards. Heavy hydrogen turned out to be very interesting. It defied the widely accepted definition of an isotope, as laid down by the father of isotope physics, Frederick Soddy.[3] Isotopes of any given element were supposed to differ from one another only in atomic weight. But heavy hydrogen behaved physically and chemically almost as though it were a different element altogether, and concentrated "heavy water" molecules containing the isotope also exhibited unique properties. This prompted Urey and his collaborators, in an unprecedented move, to give the isotope a name: deuterium.

The discovery of deuterium—regarded by one commentator as "the *bonne bouche* of inorganic chemistry"[4]—would ultimately affect research in physics, chemistry, biology, and medicine. It would also help pave the way for the atomic age. On a more personal level, the discovery changed the little family's life for the better, setting them firmly in the middle class. They were by no means rich in 1931; to afford their apartment in Manhattan's Upper West Side, they rented a room to a vaudeville singer.[5] Harold had no money but his Columbia salary, having come from poor beginnings, and Frieda had not worked since marrying. By 1934, Urey's discovery had won him the Nobel Prize. In addition to the money that came with this honor, he had also been promoted to full professor. The family soon moved to a new home in Leonia, New Jersey, where it grew with the addition of another daughter and a son. They even had enough money to hire a young African American woman from Georgia, Sadie Sherman, as live-in help.[6] For the Ureys, it was the "American dream" come true in the middle of the Great Depression.

The press took notice of Urey's success and decided that America had arrived, too. Prior to the announcement that Urey would receive the Nobel, the British physicists Ernest Rutherford and Francis Aston congratulated him on being a "brave experimenter" and remarked on how quickly he and his American colleagues had proceeded in researching this new form of hydrogen.[7] The *New York Times* took this as acknowledgment from abroad that Americans were beginning to make a larger imprint on physical research. This was, the *Times* concluded, "a return of bread cast upon the academic waters"; the American men of science (as the profession was then defined as masculine) had wrested scientific greatness from their European mentors.[8] The *Times* also took the opportunity to celebrate the fact that this new generation of scientists was working against the perception that the American intellect was solely concerned with practical matters of production and profit: "In an era when the United States is looked upon abroad as the land of materialism, the place where only the profit-making motive counts, it is good to read Lord Rutherford's words and to realize that not only the spirit of scientific research, but the ability to carry on the work of the great, lies within our laboratories."[9] As Urey's case illustrated, in some areas of science that had traditionally been dominated by European schools—especially after the devastation of World War I—the United States was actually becoming the preferred place to train.[10]

He was neither the first American scientist, nor even the first American chemist, to win the Nobel Prize. Two American chemists, Theodore Richards and Irving Langmuir, the physicists Robert A. Millikan and Arthur H. Compton, and the biologist Thomas H. Morgan, had already been honored. Still, observers saw Urey's achievement as particularly symbolic—he was one of the first chemists of international renown who had been trained almost entirely in the United States by American talent.[11] Coming from a poor background, he had risen with no access to the elite educational institutions of the East Coast. Instead he had been educated in the rural one-room schoolhouses and public schools of Indiana and in state universities in Montana and California. As his students would later write of him, he represented a new breed: "the native American scientist inspired by the problems of pure science, working not toward practical applications, but attempting to formulate the natural laws of the universe."[12]

Urey seemed a torchbearer for a new generation of homegrown scientists. By the *New York Times*' assessment, he and his less famous contemporaries signified nothing less than the rebirth of science in the New World; the young members of the American scientific elect, who made their homes in newly founded institutions of science in New York, Chicago, Berkeley, and Pasadena, were "pioneers who [gave] an impetus to physical science greater even than that which it felt in the romantic days of Faraday, Maxwell, Kelvin, Liebig and von Helmholtz."[13] The American physicist Karl T. Compton, in an assessment of American science that used Urey as its primary example, drew on this same pioneer metaphor when he claimed that, "while geographical frontiers have shrunk, the boundaries of science are wider than ever before, with more areas for

FIGURE 1 Urey being filmed in his office at the University of California, San Diego. Urey was the most eminent scientific advocate for NASA's lunar science program. Courtesy of the Mandeville Special Collections Department, University of California, San Diego.

exploration."[14] The scientists were the inheritors of a great European tradition, but also the American pioneer spirit.

OR SO THE STORY GOES . . .

Given the significance that American commentators placed on Urey's success, it is fitting that the discovery that won him his fame took place on that very American holiday, Thanksgiving. And, just as the story of that day is more complicated than it at first appears, so too is the story of Harold C. Urey. His life and career put him at the center of the most significant scientific moments of the twentieth century; he garnered science's highest honors; and he became a symbol and a spokesman for American scientific authority. He studied quantum physics in Copenhagen with Niels Bohr, did groundbreaking work with isotopes at Columbia University, ran one of the Manhattan Project's uranium isotope separation laboratories during World War II, moved to the University of Chicago, argued alongside Albert Einstein for the control of atomic weapons, helped found an institute of nuclear studies, transformed himself into an isotope geochemist, helped a graduate student perform the first successful experiment on the origin of life, and took on the origin of the Moon and planets in NASA's space science program. But who was he?

There was a public version of Urey. After winning the Nobel, it was difficult for him to escape the public eye. Not only were excerpts from his public addresses reprinted in the *New York Times* and other papers of record around the country, but the press also reported on personal events such as the births of his children. Especially within New York City, Urey was a scientific celebrity (he was certainly not as famous as Albert Einstein, who had emigrated to the United States in 1933; he did seem to possess a less enigmatic persona). Quickly, a public image emerged of Urey as an all-American. Along the lines of a scientific Horatio Alger story, this image was based on the narrative of Urey's journey from small-town boy to scientific star. Urey's reputation was that of a smart man, not a genius or even an intellectual. His colleagues credited him with a tenacious character and an uncanny ability to concentrate and apply his work ethic to the most complicated scientific or political problems. He himself insisted, "I'm not a genius and I'm not to be compared to Einstein. . . . My success came from hard work and luck."[15] Throughout his career, this image of a man who had pulled himself up by his bootstraps within the scientific community would allow Urey to speak uncondescendingly to a wide variety of audiences. It would also help protect him against Cold War critics who questioned his loyalty as an atomic crusader.

Urey felt quite comfortable with this narrative—he could paint himself as having emerged from "among the ordinary, common people of the United States." He cited his story as evidence for the claim that young people from all walks of life should be given the educational opportunities that they needed.[16] This rags-to-riches narrative did emphasize certain factual elements of Urey's past. He was born in a small town in northeastern Indiana, and his father died young and left the family to struggle on a series of unproductive farms. He grew up in desperate poverty, and he often went to bed hungry. He worked incredibly hard to make it through school and his scientific training with no financial support. In short, Urey's star did indeed rise from very humble beginnings. The subtitle of a children's biography dubbed him "the man who explored from Earth to the Moon," but it is tempting to say that in fact he traveled further—that he catapulted himself from a sweltering onion field in Indiana to the lunar maria.

But while this may seem like an all-American story, his Americanness was something Urey never took for granted; especially in his early life and career, it was something he had to claim. While the ideal of scientific objectivity is commonly understood to mean that the identity of the observer/experimenter is irrelevant to the phenomena he or she discovers, science is in fact a social activity performed by a community that has not always been open to all comers. Not all observers/experimenters have been considered equal; racialized or gendered bodies and minds were defined as not possessing the cold reason and emotional detachment that the scientific method or objectivity were understood to require. Protestant white masculinity was the mostly unarticulated standard by which the Euro-American scientific community judged prospective practitioners. As in other areas of intellectual life, the scientific community attempted to exclude, silence, and marginalize the voices of women and minorities. Although exceptional women and minorities did find paths in science, including support from some white male scientists along the way, the scientific community as a whole adopted and reinforced acceptable forms of white masculinity within its own ranks.[17]

As a white man, Urey could automatically claim at least some of the privilege afforded to his peers. However, there were parts of his identity that made him self-conscious from an early age. Throughout much of his career, Urey distanced his public persona from his religious upbringing. He was born into the German Baptist Brethren church (known more popularly as "Dunkers," for their practice of completely immersing adult parishioners in water three times during baptism), and was raised in a family in which the men traditionally became ministers in the church. While Urey did identify his father in his 1934 Nobel laureate profile as the "Rev. Samuel Clayton Urey," he did not specify the family's religion. While he emphasized the claim that he was descended from the pioneers who settled Indiana, he did not mention that those pioneers were themselves the descendants of Pennsylvania Dutch colonists.[18] Even in situations in which he made a point of drawing attention to the role that religion had played in his youth, Urey's descriptions of his religious life were vague. He tended to present himself as having had a generic Christian upbringing. He avoided references to the highly specific—or "peculiar" (as the Brethren described themselves)—aspects of his former life in the sect.[19] Urey's colleagues, in the biographical memoirs they produced after his death in 1981, likewise glossed over his childhood and his life before his graduate student days at Berkeley.[20] On the rare occasions when Urey's religion was invoked, it was only to illustrate his lifelong commitment to pacifism and his abhorrence of war.[21]

Beginning early in his development as a scientist, Urey concealed his Brethren upbringing and actively fashioned what he perceived to be an acceptable scientific identity. Partly he accomplished this simply by internalizing the values and mimicking the mannerisms of his professors; in this way, he was perhaps no different than most young scientists seeking the approval of their mentors. But he also went to great lengths to shed his Brethren characteristics—practicing pronunciation from a dictionary in his spare time to shed his accent, and joining the most Anglophilic fraternity on campus. By the time Dr. Harold C. Urey arrived in Copenhagen in 1923, and certainly by the time he married Frieda in 1926, this persona was firmly established. The words "we've arrived" were as much a proclamation of his new self's establishment as they were a celebration of the rewards to come.

A COLD WAR CRUSADER

As stable as Urey's scientific self became, it nonetheless was thrown into crisis by the events of the second half of the twentieth century. He seems to have lived rather happily with the self he had fashioned up until World War II. For the first decade of his fame as a Nobel Prize–winning chemist, not only did he never mention his religious upbringing, he went out of his way to present himself as an atheist and a scientific optimist who saw no place for religion in the modern world. He was sure that science, given the opportunity, would sweep away old superstitions and improve the world. But his role in the Manhattan Project drove him to a nervous breakdown and gave him reason to be suspicious of the "Big Science" that dominated the postwar landscape. While he was not among those scientists who opposed the use of atomic weapons, after the war was over he became very anxious about their uncontrolled presence in the world. He was subsequently demoralized by the failure of scientists like himself to influence the governance of these new weapons, his investigation by the FBI, and attacks against his character and loyalty in the popular press and on the floor of Congress.

Crisis led to a pivot. Urey transformed his postwar research program first into a study of geology and earth history, introducing new methods into the earth sciences; and then into a study of solar system formation. He also became a Cold War champion for religious conviction. While he never claimed to worship anything other than the universe itself, he nonetheless argued among his colleagues that it was only those with true "religious courage" who were willing to stand up against McCarthyism.[22] Viewing the corruption and chaos in the world around him, he wondered whether it was not daily family worship—which had played such a strong role in his own upbringing—that was lacking. In his public speeches, he insisted that it was the traditional moral teachings of the Western religions that would save the world from nuclear devastation. "It would be tragic," he said in 1956, "if science gave man the greatest view of the universe that he has ever had and destroyed the effectiveness of the teachings of our great religions."[23]

Urey found hope in religion. He argued that the language of the miraculous might be replaced by that of the magnificent. A "new prophet who [could] accept the facts of science and at the same time . . . give inspiration to fill this great void"[24] might be able to "make use of the magnificent view of the universe supplied by science and the materialistic necessities and luxury supplied by its applications to give us a sound moral life and noble aspirations."[25] Not surprisingly, the moral teachings he advocated were those that had been central to his childhood religion. He adopted the view held by many educated Brethren of the early twentieth century—that the Scriptures, and particularly the Ten Commandments from the Hebrew Bible and the Sermon on the Mount from the New Testament, were responsible for civilizing the Western world, making human progress and science possible.

Urey's newfound religious advocacy, though not a turn to religion per se, was nonetheless a reclaiming of his past. It was also, in its insistence that the moral truths of religion must meet the physical truths of science, an attempt to reconcile the disparate parts of his life story. He did this work publicly. Although he had mostly tried to avoid Big Science after the war, using contracts to support only a small research group, he chose one of the biggest scientific projects of the Cold War as the stage for this reconciliation. The National Aeronautics and Space Administration's (NASA) Apollo lunar exploration program was where Urey made his final scientific stand. He allowed NASA to use his theory of the Moon's origin along with his reputation as a Nobel laureate to shore up the scientific credibility of the costly program.[26] In return, Urey hoped to use the Moon's origin story—which he believed was connected to the very early history of the solar system—as a public display of science's ability to show humankind their place in an awe-inspiring universe.

Had Apollo provided evidence to support his theory of the Moon, we might still be talking about Urey today. We might even know his philosophy of science and religion. But Urey's Moon died with Apollo's results, and, not long after Apollo ended, Urey's reputation and popularity largely died with him. In this sense, Urey's intervention in lunar science, and his attempt to make the Moon a stage for the public unfolding of an inspiring scientific narrative that his "new prophet" could infuse with moral teachings, were both failures. Indeed, in his last years Urey deeply regretted that these failures marked the end of his career and, in his mind, led some in the scientific community to brush him aside.

UREY, BIOGRAPHY, AND THE HISTORY OF SCIENCE

In the 1960s, historians, sociologists, and writers attempted to make sense of the ascendance of American science in the twentieth century, the development of quantum physics, the atomic bomb, and the science of the Moon. Many of them showed up at Harold C. Urey's office with notepads and tape recorders, eager to ask him about his life and career. In 1963, Daniel Kevles, a Princeton University graduate student at the time, made the trip from New Jersey to California to interview the septuagenarian chemist about his discovery of deuterium, his 1934 Nobel Prize in chemistry, and his later role in the Manhattan Project.[27] In that same year, Harriet Zuckerman, then a Columbia University graduate student and protégé of the sociologist Robert K. Merton, visited Urey and collected his "reminiscences" for her study of American Nobel laureates.[28] Only one year after being visited by Kevles and Zuckerman, Urey was interviewed by the historian of physics John L. Heilbron, primarily about his early career.[29] One year after this, the journalist Stephane Groueff spent an hour asking Urey questions about his wartime research on gaseous diffusion for the Manhattan Project.[30] Before the 1960s were over, one final researcher, Ian I. Mitroff, came to interview him about his career and his views on lunar exploration.[31]

These interviewers were not the first to chronicle Urey's story. By this point in his life he had already been interviewed by a slew of journalists and researchers interested in the scientific enterprise in America. He recommended to Zuckerman that she consult the occupational psychologist Anne Roe's monograph from a decade earlier, *The Making of a Scientist*, which he felt had described well the elements of his early life that had contributed to his later eminence.[32] He drew particular attention to Roe's emphasis on the role that tragedy and hardship could play in the development of a scientist. He told Zuckerman that he fit the profile Roe described perfectly: "[She] concluded that well-known scientists are the eldest, there has been tragedy in their lives. . . . I was the eldest. My father died when I was six, and left the family in great poverty. Mother married a second time, and so forth. We're likely to be the sons of schoolteachers or preachers or something like this. My father was a schoolteacher and a lay preacher, and so forth. Right on the line."[33] Urey must have made similar comments to Kevles, who reproduced one part of this interpretation in his seminal monograph *The Physicists* when he wrote that "Urey lost his father, a farmer and minister of the Brethren Church in Walkerton, Indiana, at age six, and his faith not many years later."[34]

Urey's life story, as collected in these interviews and incorporated into the books and articles the interviewers produced, became a part of the late twentieth-century understanding of American science. These interviewers spent little time asking Urey about his childhood—a subject about which even his closest colleagues could not get him to say much. "It was hard to learn much about his early life by talking to him," remembered his collaborator in cosmochemistry, James Arnold; Urey "did not look back."[35] And so they got few details. Kevles, for example, misrepresented Urey's religion as the Brethren Church, when in fact this is a distinct group from the Church of the Brethren. And Kevles said little about the religion. Following Urey's lead, Kevles implied that Urey's loss of faith and subsequent development of a "secular brand of faith" in science was an intellectual and psychological move motivated primarily by the very tragic loss of his father at an early age.

The historiography to which these previous scholars contributed told the story of twentieth-century science in a way that reaffirmed science's role in secularizing society. This dominant view held that science has its own naturally derived set of morals, values, and norms. This was the ideal promoted by some of the most prominent scientists of the twentieth century to pave the way for a new multiculturalism and to constitute research as "pure," shoring up science's Cold War moral economy.[36] Recent historical work has given us a multidimensional view of Cold War science. We have learned much about the emergence of Cold War institutions, the transition to Cold War liberalism among scientists, the paradoxical nature of science's ability to speak truth to power, and the fate of some of the Cold War's most heroic, tragic, and in some cases enigmatic figures.[37] These histories have given us an understanding of how the dominant view came into being, and how it worked within the Cold War context.

There is another dimension to the story of Cold War secularism that most historians have not yet examined: the story of those scientists—many of whom worked within the same institutions where the secularist view was forged—who did not conform to it.[38] Working in the shadow of the bomb he helped create, Harold C. Urey rejected the notion that a secular society could survive without the maintenance of religion in some form as a source for the morals that science and technology alone could not provide. His views were in many ways very secular, and he held no beliefs in miracles or a personal deity; but he feared the materialism that might come from the disappearance of religion. He struggled during the final years of his career to promote a reconciliation of science with traditional Christian ethics. His story not only helps us trace the complex historical relationship between science and religion in the twentieth century, but it also illustrates how these complexities spilled over into the early days of space science. It therefore illuminates what was at stake, at least for some, in the Big Science of the last century.

Biography has proven itself to be a valuable historical tool. When it provides more than an uncritical celebration, the study of a life moves us "beyond easy platitudes to engage in what Clifford Geertz famously called 'thick description.'"[39] Employing biography, researchers can follow actors in and out of the networks, social movements, institutions, projects, and politics within which they were simultaneously situated. According to historian of science Michael Gordin, this becomes possible when we allow our subjects to emerge within their contexts as the heterogeneous selves they are—part and parcel, but not the center of, social and political currents. He suggests that the historian should think of the subject "as a packet of tracer dye in a turbulent stream, and then concentrate on what the consequent patterns can tell us about the stream rather than the dye."[40] This approach allows the historian to achieve the complementary goals of "deploy[ing] the individual in the study of the world outside that individual and . . . explor[ing] how the private informs the public and vice versa."[41] The subject becomes both the focal and vantage points of the narrative.[42]

The biographical approach has much to contribute to the historical study of science and religion. Recent biographical work in the history of science and religion has suggested that when historians examine "the shape science and religion take when meeting in the biographies of scientists," such studies reveal "something about the formative influences of religious background and belief on the ambitions, loyalties, and moral choices that mold scientists' lives, and even on their predilections for certain subjects and theories."[43] Along these lines, Nicholas Rupke suggests that the scientist's worldview is not synthesized out of thin air, but is formed within the context of his or her life. He suggests that we think of biography in terms of "life geography," borrowing the phrase from David Livingstone, whose conceptualization of this geography insists that we "[take] seriously the spaces in which people enact and narrate their own lives."[44] We might interpret these spaces broadly to include not only the physical spaces of institutions, cities, and nations, but also the more abstract spaces of social and political context. Biography allows us to interrogate the lives of individuals, and then to analyze how much the relationship between science and religion depends on the networks, communities, and social movements within which these individuals moved.

In this biography, I focus on the construction of Urey's identity as a scientist, his successes within the scientific community, his efforts to use his identity and celebrity to intervene in Cold War politics, and his attempt to reconcile science and religion through his public speeches and his work with NASA. The early chapters, which follow Urey through his upbringing, education, and pre–Nobel Prize career, are primarily concerned with the work he did to construct his persona and his authority. The external world of science and politics motivated and shaped Urey's self-fashioning, and in turn the self he fashioned helped him navigate these worlds. I also focus on the social and intellectual resources that shaped his worldview—including his religion. Following Urey as he traversed the landscape of American science in the twentieth century allows me to focus on how Urey and his generation of scientists saw themselves and their world. Like the "tracer dye" Gordin described, I use Urey's movement from farm boy to scientific celebrity to examine the changes in the American social and scientific landscape that made this trajectory possible.

The second half of this biography focuses on Urey's attempts to use his scientific persona and celebrity to intervene in political, social, and scientific matters. At times he had great success, as in his development of instruments and methods in the new fields of isotope geo- and cosmochemistry. Other times he failed, as when he took up the cause of world governance of atomic weapons, or when he tried to convince fellow planetary scientists that the Moon was a captured remnant of the early solar system. I have treated the successes and failures equally in this biography, as both illustrate the larger motifs of this book.

Two dominant themes emerge in this story. The first, already explained above, is that of science and religion. While often addressed philosophically in other studies, here it becomes intertwined with issues of class and race. The relationship between religion and science was not something that Urey merely thought or spoke about, but something that he lived and navigated. It was an issue of identity, having been born into an ethnic German religious sect that considered itself separate from the rest of America for most of its history. Urey fled this separateness as much as he fled the literalist tendencies of his fellow Dunkers. He also feared the inferior "otherness" often imposed by non-Brethren on members of the sect. Moving away from his childhood religion, he moved toward the naturalistic view of progress his white Protestant professors and colleagues held. And when he later revived the religious teachings of his youth, he did so under the guise of a universal Judeo-Christianity, in an attempt to create a worldview that he could share with his fellow Americans.

Urey's lived experience of science and religion is his own, but it is not wholly unique. During this period of urbanization and growth of the scientific enterprise, he certainly was not alone in having come to science from small-town, religious America. Urey and others in his generation brought with them into the scientific community the values they had learned at home. Even if they rejected the literalism or provincialism of the belief systems they were born into in favor of a more flexible, critical, or even agnostic interpretation of religion, their worldviews were nonetheless shaped by their early experiences, and their interpretations of modernity were likewise affected. It would be a mistake to assume that their encounter with scientific rationality somehow washed them clean of their pasts, as much as to believe that scientific modernity swept religion and superstition from the earth.

Urey's own approach to science and religion invites a "more sensitive historiography" that is subservient neither "to the triumphalist rhetoric of scientific rationalism nor to religious apologetics" for which science studies scholars have recently been pushing.[45] Urey's view also seems to confirm for the twentieth century what John Hedley Brooke and other historians have argued convincingly about "science in theistic contexts" in earlier centuries: that claims of secularity must not be taken at face value, since these secular visions often contain elements of the sacred in one form or another.[46] (Urey's view of scientific practice—which included the sublimation of the interests of the scientists to the power of the laws of nature—certainly contained strong elements of the sacred.)

The embodied relationship between science and religion is marked by great complexity.[47] Brooke asks that we adopt a subtle approach, "to recognize that religious beliefs and practices can shape worldviews, that worldviews may find expression in a commitment to metaphysical principles that govern theory construction, and that these, in turn, may govern the degree of assent one might give to particular explanatory theories."[48] This may in fact help explain Urey's attachment to a particular version of the Moon's history—the worldview to which he subscribed demanded an inspiring story of the solar system's origins, and his theory of the Moon put him in a position to unfold this story.

The second major theme is the peculiar nature of Cold War science and politics. While it is difficult to separate this theme from the former—it was politics, after all, and the anxiety produced by his failed political intervention that pushed Urey to take a religious stance in the Cold War—it is distinct enough to merit its own consideration. Urey's research program in the earth and planetary sciences was very much shaped by the Cold War. While he avoided weapons work, and even abandoned the isotope separation work that had made him famous, he nonetheless relied on emergent Cold War funding agencies and shaped his research program around their interests in defense and national security priorities. Like other physical scientists at the end of the war, Urey made great inroads into other fields applying physical methods and instruments to problems that had traditionally been approached observationally. In earth science, and later planetary science, the Cold War and the Big Science funding model brought new opportunities for scientists like Urey. The Cold War necessitated a better understanding of isotopes in nature. Contract research restructured laboratories and their relationships with their patrons. These forces also put the Moon on Urey's horizon, setting the stage for his "new prophet" and his final act. The Cold War also brought its own limitations. The politics of science and its relationship to the state became difficult for Urey to navigate. The scale and bureaucratic nature of Big Science worked against him; Urey's style of scientific intervention, his use of his celebrity as a source of authority, did not work in the technocratic context of NASA's lunar exploration program. He found himself surpassed by younger, more energetic scientists who better understood NASA's political priorities and were willing to fit their scientific work with them.

Comparing Urey's career before and after World War II, including his views on science and religion during these periods, is instructive. His life story is a necessary reminder that the Cold War was not an isolated period in history. Senior scientists like Urey brought to this period politics, worldviews, and scientific modus operandi that were developed in the Progressive and Depression eras, and further forged in the heat of the two world wars. Indeed, the story of how a farm boy from Indiana made his claim on the Moon can connect these periods to form the larger arc of the so-called American Century. As much as Urey's story was mythologized, it did in fact begin in a very unlikely place, spanned almost the entire century, and followed a path that touched many of the biggest moments of American science during that time.

CHAPTER ONE

From Farm Boy to Wartime Chemist

In spring 1907, 14-year-old Harold Urey stood before his fellow grade schoolers and their assembled family members in a small Amish schoolhouse in DeKalb County, Indiana. Each of the graduating students was to give a short speech for the occasion. The topic Harold chose was "perseverance," presenting "an object lesson [on] what each one of the graduates have had to do to successfully pass their examinations and reach the point of honor that they have on this occasion."[1] Instruction at the school had been basic, and yet Harold had struggled through, barely passing the graduation exam required of Indiana grade schoolers at the time. He finished at the bottom of his class, with only one boy below him in a class of thirteen students. Few of his classmates had to overcome as many obstacles as Harold on their way to graduation: his father's illness, madness, and death, and his own hard labor and poverty. It is safe to say that Harold understood perseverance.

When not in school, Harold spent most of his time working on his stepfather's modest onion farm—a forty-acre parcel of land near Cedar Lake, deep in the northern Indiana countryside—and helping to keep his family fed. Martin Alva Long, or simply "Alva," had been a hired man on Harold's grandmother Elizabeth Urey's farm in Corunna, Indiana, where Harold was raised from ages 6 to 11, along with his younger brother Clarence and sister Martha. Harold's mother, Cora Rebecca (Reinoehl) Urey, a widow, had married Alva in 1903. After the death of Elizabeth and the sale of the Urey farm, Alva had moved the family to the Cedar Lake farm.[2] In addition to Harold and his two siblings, Cora and Alva soon had children of their own.

The family lived in rural poverty. As Martha described it, the farm "consisted of some low land down by a creek that emptied into the lake. There was some higher ground for pasture and other crops, a truck patch, and an orchard. The house and barn were on top of a hill, and both were made of logs."[3] The house lacked indoor plumbing, and the summer kitchen lacked walls.

Harold and Clarence farmed on weekends and summer days. Despite the family's continued economic hardship, in later life Harold described this period as an ideal country boyhood, remembering that he slept in the attic of the log house, fished and swam in the lake, and weeded onions with his brother and stepfather during the summers: "It was a very pleasant life on the whole—terribly hot in the summertime, however, in the onion field."[4] With little extra money to buy food, the growing family might have gone hungry had it not been for the bass and bluegill that Alva and the boys were able to catch from the creek. They would eat their fill for supper, and Cora would salt the remaining fish for winter.

It was a simple country life, filled with toil and religious observance. Alva was a minister, as had been Harold's biological father, Samuel Clayton Urey. Cora's marriage to Samuel on Christmas Eve, 1891, had estranged her from her family. Cora's parents were prominent members of the local German Evangelical Lutheran Church, while Samuel was a "Dunker." He had been born into the German Baptist Brethren church, an Anabaptist sect founded during the German Radical Pietism movement. Along with the Lutherans, Mennonites, and Amish, they were among the eighty thousand German immigrants who joined Penn's Quaker colony before the American Revolution and who collectively came to be known as the Pennsylvania Dutch.[5] The Brethren insisted on a simple agrarian life and the practice of what they called "primitive Christianity." Urey's student Stanley Miller would later mistakenly assert that Urey lived a life marked by "rugged individualism," alluding to his time spent "pitching hay on his father's farm."[6] In fact, Urey grew up in a religion that valued communalism and insisted against individualism within its ranks.

Wealth and ostentation were frowned upon, and Samuel's family could be accused of neither sin. They lived in a modest log cabin on one of the smaller tracts in Fairfield Center, Indiana.[7] Samuel knew that his religion was a source of friction between his family and Cora's, and so he never asked her to convert. Nonetheless, she chose to become a Dunker, despite the fact that doing so meant "no jewelry, no puffed hair, and no fancy clothes," and few of the comforts she had known while living with her more prosperous family.[8] Cora's Dunker conversion was sincere; even after Samuel's death, she remained a devout member of the Brethren community.

The family's daily life, as Harold later recounted, was shaped by their Dunker faith. Before the day's work could be done, mornings began with prayer and family worship.[9] They observed a strictly modest dress code, a "Gospel plainness" that their religion required.[10] On Sundays, regardless of the weather, the family drove their horse and buggy five miles to the Dunker meetinghouse in Cedar Creek. Harold later recounted:

> In this church, and its Sunday School, I learned my knowledge of the Bible mostly. . . . In the old days, the women and men sat in different sections of the church. The women wore long skirts and long sleeves with high-necked collars, a marked contrast to the clothes of today. The men wore coats with no place for neckties. The sermons were about ½ hour long, and my father and stepfather both preached in this church. The communions and the full-scale meal in honor of the Lord's Supper, where we washed each other's feet, took place once a year.[11]

The majority of Brethren that Harold grew up around in the Corunna area would have held that the Bible was the only guide to living as a Christian. What the Bible advocated was a life of discipleship based on a strict commitment to the Ten Commandments, as interpreted by Christ in the Sermon on the Mount. (While Urey would claim to have rejected the literalism of his religion as a teenager, these chapters of the Gospel of Matthew remained his lifelong favorites.) The Brethren also held then the view that the Bible was an infallible record of human history, and that all great advancements in society were due to the Bible's influence.[12] In the words of the Brethren ministers Owen Opperman and Charles M. Yearout—words that seem as though they could have inspired Urey's later position—the Brethren agreed that all the "comforts and learning and great improvements" of America were produced by the Bible's impact on society: it had elevated humankind from "a state of barbarism and superstition to a high plane of morality and enlightenment."[13]

Harold had little memory of Samuel, whose time as the family patriarch had been marked by illness, and ultimately culminated in his death when Harold was only 6. He knew that his father had encouraged his education: “He was able to teach me to write a bit, and he always insisted that I should bring books home at night and read a small amount to him.”[14] In Harold’s account of his life, Samuel was a dedicated father and educator.

Samuel had been an educator, although his aspirations in this arena were cut short by his illness. Shortly after their marriage, Samuel had moved Cora to Walkerton, Indiana, where he had taken a job as a school superintendent. He had put himself through a bit of college and had taught in the small schoolhouses of Fairfield Center, and the job in Walkerton was an attempt to better his prospects. Less than a year into his new appointment, on April 29, 1893, Harold was born. Aside from the birth of their first son, Samuel and Cora’s time in Walkerton was far from joyful. They were barely settled in before illness forced them to move again. Tuberculosis had struck two of Samuel’s sisters, and he moved his small family back to Fairfield Center to help. He soon contracted the disease himself and moved the family west on a doctor’s recommendation that the dry climate might improve his health.[15]

What Harold did remember of his father was from the brief time the family spent in Glendora, California. In 1897 Samuel took up the life of a Brethren missionary minister, and the little family—which now included a second son, Clarence—joined the ranks of the more than six hundred Brethren that the minister Matthew Mays Eshelman brought to Southern California to settle near his new college at Lordsburg.[16] It was during this period that young Harold began to observe and recognize his family’s hardship. Some of these memories were sensory—the sweet taste of a small spoonful of mission fig given to him by his mother on the front porch. Others were more emotionally fraught—when an interviewer later asked him about his childhood, he summed it up succinctly: “Poor. We were very poor. I remember terribly poverty stricken days.”[17]

As a missionary minister, Samuel preached alongside Eshelman, serving congregations in the towns of Covina and Colton.[18] The position was not salaried, and so he found work in a local packinghouse, hammering together wooden strawberry crates. The manual labor exacerbated Samuel’s tuberculosis. One of Harold’s nieces later recounted the family’s story, telling of how Samuel “hammered and coughed, hammered and rested, as his little boys, Harold and Clarence, played in and around the

FIGURE 2 Samuel Clayton Urey and Cora Rebecca Reinoehl, wedding photo, 1891. Courtesy of the Urey/Cullen family.

workshop."[19] The family was destitute, surviving partly on the charity of their Brethren neighbors.

As bad as things were, they soon worsened. Samuel's mental health was rapidly declining. He knew that he was dying and that he was leaving his wife, newly pregnant with their third child, to raise the family alone and in dire poverty. Nothing could console him. As family member Kay Waters put it: "Neither prayer nor love could reverse his condition. . . . It is no surprise that this responsible, devout, loving man, unable to support his family, knowing he was to die, went briefly mad."[20] This mental decline was perhaps accelerated by Samuel's self-imposed regimen of fasting.[21] On the Fourth of July, 1897, after he had finished preaching an evening sermon in the Colton church, Samuel suffered what his family later described as a complete emotional breakdown. A newspaper report from the time suggests that in a fit of madness Samuel attempted "to kill

FIGURE 3 Samuel, Clarence, Harold, and Cora Urey, family portrait, 1895. Courtesy of the Urey/Cullen family.

FIGURE 4 The Urey family in the strawberry patch, orange trees in the distance. California, 1897. Courtesy of the Urey/Cullen family.

FIGURE 5 Samuel Urey hammering packing crates, California, 1897. Harold and Clarence play on the packinghouse floor. Courtesy of the Urey/Cullen family.

an old man at Lordsburg" that evening.[22] A county judge committed him to the Southern California State Asylum for the Insane and Inebriates in San Bernardino, where doctors determined him to be "acutely insane," and suffering from tuberculosis meningitis. The disease had reached his brain.[23] He remained institutionalized within the asylum for eight months.

Restoration would not come. Samuel was released in March 1898, and two months later he traveled to his mother's farm to sit at his sister Etta May's deathbed. She died that August. Soon after Etta May's death, Samuel abandoned all hope that the California climate would improve his health. If he was to die, it was better to die in Indiana, where Cora and the children would be near family. In June 1899, they moved back to his mother's farm.[24] Less than a half-year later, Samuel died, leaving Cora and their three children in the care of his mother. "He was a man of a brilliant mind," his obituary read. "The Brotherhood at large has sustained a great loss in his death."[25]

For her part, Cora seems to have shouldered the burden of raising her children in the shadow of illness and poverty with little complaint. The family records describe her as "truly a strong, hardworking and brave pioneer mother and wife,"[26] and there is nothing to contradict this description. During Samuel's incarceration, Cora supported herself and her boys as a washerwoman. Even while several months pregnant with her daughter, Martha, Cora bent over the washtubs or bathtubs of her employers, "turning out freshly ironed clothes whose white smoothness belied her knuckles made raw on the washboard."[27] "I have always thought I must have been the most unwanted child in the world," Martha later wrote. "How could anyone want a baby at a time like that? Yet, as I grew up I felt loved and extra special. My dear wonderful mother was a remarkable woman."[28] Harold, more reserved than his sister about such matters, nonetheless felt much the same—Cora was his moral compass. In a speech accepting the first of many accolades in his career, he credited his mother as having been one of his most influential teachers: "It was she who taught me that 'man does not live by bread alone but by every word that proceedeth out of the mouth of God.' Of all the lessons that I have learned in my life, this one has been most valuable."[29]

A NEW LIFE

Life on Alva's Cedar Lake farm was free from the specter of death that had colored Harold's early childhood, but was still demanding. Brethren families in agricultural communities at the turn of the century typically remained isolated from the non-Brethren world.[30] Harold's family, as he often emphasized, was pioneer stock, having settled the wilderness of Indiana. The frontier lifestyle persisted within the Urey household, since they had little money. As it had for previous generations of Brethren, life revolved mostly around the production from raw materials of all the necessities of life—including clothing and food. As a result, each member of the family was an essential part of "an industrial, social, moral and religious organization."[31] The home was the center of religious life for a Brethren family, and home life included a daily routine of scripture reading and worship. "The home without daily family worship," wrote one Brethren sociologist, "was considered to be without true Christianity."[32]

There was no high school in the Cedar Lake area, however, and this afforded Harold the opportunity to leave. Money from Samuel's life insurance policy went toward room and board with relatives on the outskirts of the small town of Kendallville, Indiana. Here he lived with his maternal grandparents and other relatives from his mother's side of the family throughout high school. Little more than a dozen miles from the family farm, Kendallville offered Harold a larger and more heterogeneous community. It was here that he would come into his own.

Urey was not completely cut off from his mother and siblings. When harsh winter weather did not prohibit it, he often rode his bicycle back to the farm on weekends. He participated in the family's religious life while home. As he was now living with Cora's family, however, he must also have been introduced to a type of family life unfamiliar to him. The family was active in Kendallville's First Evangelical Church; while devoted Christians, they did not observe the same types of restrictions in dress or behavior as did the Brethren. Harold likely also ate better while living with the Reinoehls. According to family sources, Martha Reinoehl "served bounteous meals every day. Breakfast was a feast which always included her favorite sugar cookies, pie, eggs, bacon or ham and much, much more."[33] Just as in Harold's mother's home, breakfast was preceded by family prayer and worship.

Harold's time away from the farm had a marked effect on him. Prior to this point, he had been educated in relatively closed and orthodox settings, in very modest schoolhouses; his social group had consisted primarily of his siblings and Sunday school classmates. In Kendallville he immediately became self-conscious about his lack of sophistication, and in hindsight considered himself a "raw youngster" compared to his new cohort: "exceedingly timid, immature and unaccustomed to a town of 5,000 people."[34] Harold's initial social anxiety did not go unnoticed by his new classmates; as one later recalled, she and her friends understood his shyness and reticence to be a product of his Amish schooling.[35] By Urey's account, at the root of his insecurity were his country mannerisms, which no doubt included his Brethren peculiarities—his plain dress and his accent; he saw these as obstacles to his peers' acceptance. He would later write to a childhood friend, "What a crude country boy I was at that time, and you treated me so well."[36]

The culture shock Harold experienced fits well with that described by other Brethren youth who attended high schools or colleges outside their sects. These educational environments were more diverse than the closely knit communities from which they came, where church life required a separation of male and female congregants. While these young Brethren reported feeling ashamed of their characteristic appearance and speech patterns, the experience could be a liberating one. Non-Brethren high schools and colleges provided them a "zone of invisibility" within which they could "experiment with questionable customs and practices without the knowledge or censure of their home community."[37] For Harold this zone was defined by the absence of his primary living role model, his mother Cora. In Kendallville he socialized with girls and attended parties. High school, and later college, became spaces within which he could reinvent himself.

Harold did his best to shed his "raw" characteristics as quickly as possible. He found that one effective tactic was to throw himself into his studies. If a country boy was naive, Harold would become erudite. Although he had barely passed the high school entrance exam, his bulldoggish tenacity, combined with a newfound love of learning, soon put him at the head of his class. "It is interesting that a country boy who barely passed the examinations out of the grade school of Indiana led his high school class immediately, and continued to lead his classes in college from then on," he noted.[38] He earned a spot on the debating team. His shyness abandoned, he became known among his classmates as a budding orator; they began calling him "Professor," a nickname that stuck with him through his college years.[39]

Urey later stated that he began moving away from the church in his teenage years after reading the works of the agnostic freethinker and orator Robert G. Ingersoll.[40] Urey does not specify which of Ingersoll's works he read, or where he encountered them. He might easily have read summaries of Ingersoll's arguments, along with Brethren rebuttals in the Brethren periodicals the *Gospel Messenger* or *Brethren at Work* before reading them firsthand; these publications regularly felt the need to defend Christianity and temperance from Ingersoll's attacks; many of his works presented a historical interpretation of the Bible that questioned its divine inspiration and thus argued against any form of literalism.[41] The Bible—like works of history, law, government, and science—was not "above and beyond the ideas, the beliefs, the customs and prejudices of its authors and the people among whom they lived."[42] The evidence he presented—that the Bible's authors were culturally situated—spoke against "one ray of light from any supernatural source."[43] Not only was the Bible mistaken with regard to the place of Earth in the heavens, the motions of the sun and planets, and most matters of "creation, astronomy, geology; about the causes of phenomena, the origin of evil and the cause of death," Ingersoll also argued that scripture was not "any nearer right in its ideas of justice, of mercy, of morality or of religion than in its conception of the sciences."[44]

Urey adopted Ingersoll's rejection of literalism, although he never went so far as to accept the argument that all of civilization, and "all that we call progress," had been accomplished in spite of the Bible.[45] On the contrary, during the Cold War, Urey would repurpose the Brethren argument that the Ten Commandments and the teachings of Jesus—especially the Sermon on the Mount—had civilized the West and allowed for the rise of science. He would turn this into an argument for the importance to science of religion's continued existence. Still, it is not difficult to imagine the great impact that the freethinkers might have had on the young Urey, given his upbringing within a pious household and an orthodox community.

Harold owed his education to his father, a man of whom his strongest lasting memories were the deathbed encouragement of his studies. Although stepfather Alva had become an essential part of the family, Harold never regarded him as a paternal influence. It is no surprise, then, that he began to gravitate toward his teachers and to seek their approval. Urey found among his teachers new male role models. This is where his transformation began, as he patterned himself after these surrogate father figures. His primary high school mentor was his Latin teacher, E. E. King, who inspired him to consider taking up a career as a Latin teacher himself. For Urey, King was just the first in a line of instructors who would come to influence his image of himself as a scholar and help him to reshape himself gradually into a cosmopolitan man of science.[46] He would look to them for guidance and acceptance; he would experiment in fashioning himself in their images as he searched for his place in the world; and he would build a unique persona from the bits and pieces taken from each man, just as he shed those bits of his previous self that no longer fit.

With Harold's poor family background, his understandable aversion to farm life, and his newfound love of education, the best career he could imagine for himself after high school was as a teacher. Science did not yet occur to him to be a suitable career; while he had done well in high school biology and physics, he found the subjects boring. After graduating in 1910, he first followed briefly in his father's footsteps and received his teaching certification in 1911 from Earlham College, a Quaker school in Richmond, Indiana, before going into the small country schoolhouses of Indiana as a schoolmaster.

In 1912 Alva attempted to improve the family's lot by moving them to an eighty-acre homestead northwest of Big Timber, Montana, which he had purchased at a low price through the Carey Act two years earlier. The act provided newly reclaimed and irrigated land to farmers who promised to make the land productive. As Harold's sister, Martha remembered it:

> Daddy had read a glowing account, in our church paper about some land out west, that sounded very inviting. So, along with a number of other church members from various parts of the country he went to look the situation over. He was pleased. There were two large reservoirs filled with water for irrigation, the ditches were dug to bring water to the ranches and he saw fields of wheat and alfalfa growing. Truly, it was a "land flowing with milk and honey." As a result we began to pull up our roots.[47]

The family packed up only the household necessities, a few farm tools, and the organ. With a pair each of horses and cattle in tow, they headed out for Montana by rail. Unfortunately for Alva, this new farm proved as unsuccessful as the last, if not more so. The soil was unproductive, the winters unbearably harsh. Despite the hard work Alva and Cora put into it, the venture failed. After less than four years, unable to pay his mortgage, he was forced to sell all his livestock and farming equipment—everything but the land—and leave.[48]

Harold had relocated to Montana to stay near his mother. The state's boosters at the turn of the century billed it as a land of opportunity; in addition to land reclamation, the state and its entrepreneurs were heavily invested in railroads, logging, and mining. Populations were springing up in undeveloped areas. Harold continued teaching, moving from the one-room country schoolhouse to the one-room frontier schoolhouse. He first taught in a small wooden building that housed fifteen to twenty students at the foot of the Absaroka Range, on the eastern edge of Yellowstone National Park. After one year, he moved across the valley into the Gallatin Mountains at the northern edge of Yellowstone, where he taught for one additional year in a mining camp and became principal of a two-room schoolhouse.[49]

THE UNIVERSITY OF MONTANA

Teaching did not earn Harold much money, and so he boarded with the families of his students in exchange for tutoring. When the boy he had been tutoring in the mining camp (a boy he did not consider exceptionally bright) left to attend college in Bozeman, Montana, Harold decided, "If perhaps I were going to get ahead in the world, I should go to college also."[50] In fall 1914, just as World War I was beginning in Europe, he enrolled at the University of Montana in Missoula.[51]

It was no easy matter for Harold to afford tuition at the university. Teaching had not allowed him to put aside much money, even living as frugally as he did. According to one of his later collaborators, he slept and studied in a tent during his first academic year, and during the summer "worked on a road gang laying railroad track in the Northwest."[52] In his account of his Montana years, Harold made no mention of living in a tent. He did remember spending his summers performing hard labor in order to cover his costs: "One summer I spent working on the railroad which was being electrified running through Missoula. I worked partly on this railroad and partly on an irrigation project up the Missoula River region."[53] He also recalled waiting tables in the girls' dormitory and, in his second year, taking a job as an assistant in the biology department.[54]

A happy coincidence brought Harold to the physical sciences at Montana, despite his high school experience. In his idolization of King, he saw himself as a man of letters—an orator and a master of erudition. But life as a teacher had shown him that he needed to think more practically about a professional life if he was to escape the poverty and hardship of his youth. He first set his sights on psychology. The Psychology Department, however, had a policy of not admitting freshmen students. Discouraged from this path, he registered instead in chemistry and biology courses—a turn of events that put him in one of the young university's most dynamic departments. The botanist and naturalist Morton J. Elrod had come to the university in 1894, one year after its founding, to head its science department. Elrod was one of the university's most active faculty members; during his tenure he built up the biology department, founded an instructional biological station and freshwater laboratory on Flathead Lake, and served as president of the Montana Academy of Sciences, Arts and Letters.

Studying in Elrod's department meant that Harold got a healthy dose of both laboratory science and natural history fieldwork. Elrod took his students into nature as often as he could—whether the surrounding mountains of the campus, Flathead Lake, or Glacier National Park. It also meant that Harold came to see women as participants in science. The biology department taught large numbers of women, and Elrod himself promoted the idea that women could play more than a supporting role in science. Elrod's wife, Emma, and his daughter, Mary, were fixtures in the department; Mary received a bachelor's in biology from the department in 1911 and later held a position as one of its instructors. One of Elrod's first and most famous students at the University of Montana, aside from Urey, was Jeannette Rankin, the first woman elected to the US Congress.

Perhaps it was Elrod's influence that led Urey to take women on as students and research assistants in his own lab, and to promote their scientific careers. American science and higher education at the turn of the century were, with few exceptions, the exclusive domains of white Protestant men. Urey's experience, attending a young university on the western frontier, consisted of some uniquely progressive elements. Elrod's insistence on having women in the field and in the laboratory, as equal partners in the scientific enterprise, was not the norm. Had Urey attended Harvard or another Ivy League school during this time (as so many of his later, more distinguished peers did), rather than a land grant university with a broader coeducational mission, his contact with women scientists would have been very different; the elite institutions of the East Coast did not open their doors to women until the second half of the twentieth century. There were women working in these elite institutions, but their contributions to science (many of which were original and significant) were not considered equal to those of their male colleagues, and their work was often made invisible.[55] This is not to say that Montana was a paradise for women scientists; clearly their advancement within the university was still limited, despite Elrod's views. However, the lack of entrenched tradition at schools like the University of Montana opened up new possibilities for this otherwise marginalized group.[56]

Urey soon became a biology major, and in 1915 he completed his first study in natural history under Elrod, "Biology of a Slough near Fort Missoula." In these early years he was particularly taken by protozoology, and he later claimed that this was the beginning of his interest in the origin of life: "I was immensely fascinated in the first days of my course in

FIGURE 6 Urey's time as a student at the University of Montana gave him the opportunity to refine his scholarly persona, and to grow into his nickname, "Professor." These yearbook photos from the University of Montana *Sentinel*, 1916–18, document these years of personal and professional development. Courtesy of Archives and Special Collections, Mansfield Library, University of Montana.

FIGURE 7 A coed biology classroom at the University of Montana. Photo by Morton Elrod. Courtesy of Archives and Special Collections, Mansfield Library, University of Montana.

biology with these small microscopic animals, and thought that in some way they represented the simplest organisms, and that perhaps the origin of life on the Earth was bound up in the study of these organisms."[57] None of the hard or menial labor that Urey performed during his first year at school seems to have detracted from his studies; in 1916 he was awarded an endowed prize after being named "the best student in the Department of Biology."[58]

The University of Montana was still in its infancy, and so the small student body was both close knit and well known to their professors. After being awarded his prize, Urey was given a post as an assistant in the department, and the biology faculty surely knew Harold. His instructors, including Elrod, soon became his close advisers. He certainly would have found in Elrod, a turn-of-the-century Renaissance man, an example of a well-rounded and erudite scientist. In addition to performing scientific work, Elrod published and spoke on a wide variety of social and cultural topics, from the advancement of women and the benefits of camping for girls, to the relationship between climate and civilization and the philosophy of death. He also read and wrote poetry and was one of Montana's most active early photographers.[59] Elrod easily became one of Harold's professional archetypes.

The instructor with whom Urey spent the most time was Archie Wilmotte Leslie Bray. Hailing from Sheffield with a Cambridge education, Bray was only ten years Harold's senior. In Urey's own apocryphal biography of Bray, he imagined the Englishman as a scientific hobo. Describing him to science writer Shirley Thomas in 1963, Urey claimed that Bray was so eager to see the United States that he had spent all his money on boat passage across the Atlantic. Having no funds upon his arrival, he "started his sightseeing by train—freight train."[60] Urey's account goes on to claim that Bray was kicked off of the train in Missoula. As he had no credentials with him, he took a job as a janitor in the university. At some point, so Urey's story went, the university realized that they had an Oxbridge biologist in their midst, and he was promoted to the position of assistant professor in zoology.[61]

Urey's biography of Bray is an almost complete fabrication. By Bray's account, he traveled through most of the United States and some of South America at the turn of the century, before he even entered Cambridge in 1905. In between graduating from Cambridge and taking up his position in Montana, Bray traveled through Newfoundland for two years, spent considerable time as a teacher in Labrador, and then spent two years after this traveling through the rest of Canada. After his time in Canada, Bray received a graduate degree in philosophy from the University of Oregon and completed some graduate coursework at the University of Montana just on the eve of Urey's arrival.[62] Starting in 1913, Bray spent one year in Montana as an instructor and three years as assistant professor. As for Urey's account of Bray as a janitor, according to an obituary, Bray held several jobs before "settling down as an educator," including cowhand, muleteer, cabin boy, hotel porter, ditchdigger, and draftsman.[63]

While Urey's biography of Bray was clearly a heroic fantasy—perhaps one that allowed him to imagine the scholar as having come from a hard-scrabble background, similar to his own—his assessment of Bray as a born educator seems to have been entirely accurate. Bray left Montana in 1917 and became a graduate student and teaching fellow at Harvard. In 1925 he joined the faculty of Rensselaer Polytechnic Institute, where he eventually became a founding member of the institute's Department of Biology. Bray distinguished himself as a generous adviser and educator throughout the rest of his career.[64] It was Bray perhaps above any other instructor who helped Urey shed the skin of the shy, self-conscious Indiana farm boy.

Bray organized a group of young men, including Harold, into a philosophical club that he called the Authentic Society. The club was modeled after the Cambridge Apostles, a "free discussion society" that boasted among its former members John Stuart Mill and Alfred, Lord Tennyson, along with other "men of world prominence."[65] During Bray's time in Cambridge he was a member of the "Authentic Club," a reorganized version of the Apostles.[66] Bray's Authentic Society became the Alpha Delta Alpha fraternity in 1915, with Harold as one of its founding members.[67]

Bray introduced the young men of Alpha Delta Alpha to the traditions of Oxbridge, and by doing so created a distinctly Anglophilic fraternity. In their living room, the young men hung a portrait of Sir Galahad. (After his death, Bray's portrait was hung side by side with the knight's.)[68] Urey's participation in the Authentic Society changed the nature of his Montana education. Bray brought Oxbridge to the frontier. In retrospect, Urey concluded that Bray had given him a traditional tutorial education, and in effect bestowed worldliness on him: "I soon realized that under him I was getting a Cambridge education and was gradually changing from a little country boy to more nearly a man of the world. . . . Professor

Bray was just a splendid, model teacher who opened up the whole fascinating world of science to me."[69]

Urey's classmates in Montana began to see the scientist in him, although they tried not to let his newfound talents go to his head. In "The Color Thief," a short story written for the 1917 yearbook, an anonymous classmate reimagined him as the main character in an Arthur B. Reeve mystery. In place of Reeve's Professor Craig Kennedy (the American Sherlock Holmes), the author introduced Professor Harold Urey, "the great scientific detective." The story clearly had some fun at Urey's expense. Not only did it play on his longtime nickname, it also had him dressing as a woman and posing as a chaperone to gather clues at the coed dance. It was a case of stolen powder and rouge, the likes of which had not yet appeared in the West. The only way to solve it was for the detective first to familiarize himself with the tastes and smells of all the locally available products, then kiss each girl on the cheek as they entered the gymnasium. The mystery was solved comically, when Urey kissed another boy, also in drag, who had stolen the makeup in order to sneak into the dance. Behind the adolescent humor, the author still paid tribute to the budding intellect and "his keen mind back of his guarded eyes."[70] Based on his yearbook photo from the same year, Urey now looked the part: well dressed, immaculately groomed, and with a visible air of confidence—very much in line with his later reputation as the best-dressed man in the laboratory. This was no shy or naive country boy.

THE IMPORTANCE OF PASSING

The yearbooks from Urey's time at the University of Montana include many inside jokes like "The Color Thief." They also performed the more serious task of defining the collective identity of the university's student body. The university in those years can hardly be said to have been racially or ethnically diverse.[71] Montanans and other westerners—who had only recently "claimed" their land from native tribes, and whose growing communities continued to dislocate and dispossess those tribes—had a highly constrained definition of Americanness built on the exclusion of the nonwhite "other," imagined to be unfit for American life. The students of the University of Montana were aware of this history, or at least one version of it. They imagined a cooperative and peaceful pioneer past for the Missoula basin, in which white settlers and native people had cooperated. Depictions of Missoula's pioneer past treated the native population nostalgically, but these depictions also celebrated the progress of white civilization that had displaced them.[72]

It was not just nonwhites in Montana whose suitability for modern life was questioned. Whiteness itself was fracturing during this time. The period between 1840 and World War I saw increasing scrutiny of what the historian Matthew Frye Jacobson has described as "the internal divisions of whiteness."[73] It was "a shift from one brand of bedrock racism to another—from the unquestioned hegemony of a unified race of 'white persons' to a contest over political 'fitness' among a now fragmented, hierarchically arranged series of distinct 'white races.'"[74] By and large this fracture of whiteness was the result of the wave of new immigrants entering the United States from Southern and Eastern Europe. One of the primary categories by which these groups were ostensibly judged (and what the pseudoscience of the day was designed to test) was their fitness for self-government. It was by this category that the true Anglo-Saxon was distinguished from the inferior European. Even German groups with colonial antecedents—which would seem to be included within the Anglo-Saxon identity—could come under scrutiny.[75] What Urey would later experience secondhand through his Jewish students and colleagues as anti-Semitism, he experienced firsthand in this context. Even progressive scientists, swayed by the promise of eugenics and entrenched in the prejudices of their time, questioned the fitness of the Brethren, their ability to participate in democracy, and their loyalty to their adopted country.

Despite having taken steps toward modernization near the end of the nineteenth century, the Brethren were seen as a problematic group within modern America. In his 1911 monograph, the eugenicist Charles Davenport singled out Amish and Dunker communities as groups that had in effect committed race suicide by erecting "the barrier of religious sect . . . again and again to insure the intermarriage of the faithful only."[76] In his role as head of the Eugenics Record Office, Davenport sent one of his fieldworkers, Amey Eaton, to Lancaster County, Pennsylvania, to study the Amish. Her study led him to the conclusion (based on Eaton's reported rate of epilepsy and other physical defects among this one sect) that the equivalent of a genetic time bomb was waiting to wipe out the entire community: "A defect is in the blood of some of the strain that in time will affect the entire sect who remain in that part of the country."[77] Although he briefly considered the claims of other social scientists (including one Brethren sociologist) that recent liberalization within the Dunkers was leading to a gradual lifting of the prohibition against marrying outside the church, he nonetheless concluded that, considering intermarriage within the congregation for several generations, "we can but regard such small sects as eugenically unfortunate."[78]

Davenport's condemnation of the Amish and their kin was directed not only at the rate of crippling birth defects within their closed populations, but also their social attitudes. The eugenicist regarded the Pennsylvania Dutch as defective due to their rejection of modern technology and their pacifism.[79] The eugenics movement was to a great extent the scientific expression of the fracturing of American whiteness.[80] Like the Irish, Jewish, and Southern and Eastern European immigrants who arrived in America at the turn of the century, the Brethren tended to fall on the wrong side of the categories drawn up to demarcate the varying degrees and divisions of whiteness during this period. That the Amish and the Dunkers seemed to reject modernity, and that they had a long history of refusing to participate in government, civil society, and the military, marked them as something less than fully white.[81] Like Henry Goddard's "feebleminded" Kallikak family, according to eugenicists, the Brethren were unfit to be granted the full privileges of whiteness.

But at the same time that university life and the pressure to fit in acted to exclude Urey, they also provided him with a model of conformity that could mask his difference and salvage his perceived whiteness. Urey set out self-consciously to shed his Brethren characteristics—dropping his accent and adopting new speech patterns—and compensated for his remaining rough edges. A bricolage of what he considered the best qualities of his instructors, mentors, and peers became the blueprint for his transformation. While his love of learning and his talents were no doubt genuine, he worked hard to cultivate not only his own intellect, but just as important, the appearance and demeanor of a man of learning. Adopting a more sophisticated appearance was not easy for Harold, who still had limited financial resources. On one trip to visit his family in Idaho for the holidays, he and his brother Clarence traded neckties; this allowed Harold to return to Montana appearing as though he had bought a new tie.[82] He adopted Bray's Oxbridge model of scholarship, and he no doubt internalized many of the social and cultural ideas he learned within the

Authentic Society and the Alpha Delta Alpha fraternity. Urey was in effect erasing his Brethren identity.

THE FRONTIER GOES TO WAR

Under Bray's direction, biology became Urey's principal love during his college years, though his coursework also emphasized chemistry. He graduated from the University of Montana in only three years, and went to work for Bray in the summer of 1917 gathering ticks in eastern Montana in a project aimed at ameliorating the area's spotted fever problem.[83] He had entered the university in 1914, just as hostilities in Europe were heating up, and he graduated just as America officially went to war. The Selective Service Act became law on May 18, 1917, and draft registration began that June. The call to arms put an abrupt end to Urey's work with Bray, who dropped his spotted fever research and, after arriving at Harvard, enlisted as a biologist in the Chemical Warfare Service (CWS) in Washington, DC.[84] Before leaving Montana, Bray advised Urey to put his scientific knowledge to use in the war effort, insisting that "a trained chemist should serve on the chemical side."[85]

Many of Urey's fraternity brothers also joined up. According to the records of Alpha Delta Alpha, thirty-seven of its forty-one members enlisted. The fraternity formed a cooperative relationship with Missoula's Student Army Training Corps (SATC), holding one tent as an unofficial fraternity headquarters within the camp during training.[86]

The peer pressure to enlist went beyond the fraternity. The entire campus community was taken with war fever. *The Sentinel* yearbook adopted a patriotic tone that had been absent in the previous years of US neutrality, and in 1919 it took on the mantle of "the university's first wartime annual." Describing the 1917 transformation of the student body, the editors wrote: "With the coming of the cloud [of war] a student body was changed overnight; transformed from a rollicking, carefree band of young men and women enjoying four years of university romance to a group of determined Americans face to face with a stern reality." In a political cartoon titled "As We Visualize Our Boys in Action," an oversize University of Montana mascot, the grizzly, is seen chasing a dachshund wearing Kaiser Wilhelm's helmet and mustache, cheered on by an American-flag-waving University of Montana coed.

FIGURE 8 The ROTC army camp on the University of Montana campus. After the United States entered World War I, many of Urey's peers and fellow fraternity members enlisted and moved into the camp. Courtesy of Archives and Special Collections, Mansfield Library, University of Montana.

The rest of the state of Montana, which surpassed all other states in enlistment rates and draft quotas, was likewise swept up in the war effort. With 12,500 volunteers and another 28,000 draftees, almost 10 percent of the entire population of Montana went to war.[87] Along with war fever came a good amount of prejudice and hysteria; the state also became a hotbed for the wartime "Americanization" movement. Montana, like many other western states, had its own version of the "genuine" American. Montanans celebrated the Frontier American, which they defined as a modern nation-builder undiluted by alien blood or influence.[88]

For Urey, this must have been an exciting but frightening time. Although he had already done much to reinvent himself and shed his Brethren characteristics, this turn of events must have made complete dissociation ever more urgent. It is not surprising, then, that while in Montana, Urey spent his spare time practicing the pronunciation of words from *Webster's Unabridged Dictionary*, attempting to keep his Brethren accent at bay—a practice he had begun in high school.[89] While many of Urey's colleagues held this up as an illustration of the self-discipline with which he raised himself by his bootstraps, it may in fact have been a survival tactic.

The period of neutrality before the war was tense, with Montana's Anglo-American population growing suspicious of the imagined German agents and agitators that wartime propaganda told them might be in their midst. By the time draft registration began, the stage was set for anti-German hysteria and a climate of intolerance producing hundreds of instances of abuse. The war amplified the already prominent Anglo-American nativism of the western states; according to the historian Frank Van Nuys, "Regardless of birthplace or background, [immigrants and ethnics in the West] were told that one was either unequivocally American or not American at all and thus deserving of the severe condemnation reserved for the disloyal."[90] Ethnic German Americans, accused of offenses ranging from poisoning American enthusiasm for the war to outright treason, were pressured to buy Liberty bonds; they saw their presses put under arbitrary scrutiny and their organizations accused of being funded by the German government. They were also subjected to humiliating, mob-induced rituals including tarring and feathering, forced marching, and flag-kissing ceremonies.[91]

Especially troublesome to "true" Americans was the perceived clannishness of some groups of ethnic Germans—particularly those who had immigrated recently or who had for generations refused assimilation. Needless to say, most "sectarian" religious Germans were lumped into this latter group. Those churches that chose to defend their German heritage became easy targets for anti-German abuse.[92] As the political and social climate grew more intolerant, ethnic Germans of all stripes suddenly faced intense pressure to assimilate to what Anglo-Americans presented as the true American culture. The historian Frank Luebke noted, "For the German-American who valued his cultural heritage this meant, at best, that he was a second-class citizen, inferior to Americans of English antecedents; at worst it meant that he was perceived as the agent of a foreign despot."[93] Although it did not free them from all suspicion, some German Americans responded to the campaign for uniformity by embracing the tenets of the emerging "superpatriotism" that defined loyalty as a subordination of everything—including ethnicity, conscience, and individuality—to the successful prosecution of the war.[94]

The Church of the Brethren was recognized by the draft board as a historic peace church, and thus Urey could have claimed conscientious objector status. However, claiming such status might have undone much of the transformation he had worked so hard to achieve, and it would potentially have put him in real danger. Claiming the status based on his affiliation with a German American peace church would have been problematic to say the least—congregants of culturally German churches in the West faced trials. German Lutherans came under fire from one federal district judge: "Instead of trying to remove the foreign life out of their souls, and to build up an American life in them, they have striven studiously from year to year, to stifle American life, and to make foreignness perpetual. That is disloyalty."[95] In Glendive, Montana, an unruly mob hanged a Mennonite nearly to death for pacifist views.[96]

Young members of the Church of the Brethren may not have felt obligated to avoid military or war-related service. The church's missionary and educational efforts at the turn of the century had made it less legalistic in its positions and more tolerant of the individual consciences of its members. As for young, draft-age Brethren like Urey who were attending secular colleges, it was common for them to feel a great allegiance with their non-Brethren fellows. This made it "harder to break with their community associates on the peace question than in the days when the church maintained an exclusive church fellowship."[97] Brethren responses to the draft were inconsistent. Some chose to train for war (a few even became officers); some took noncombatant service as conscientious objectors; and some refused service altogether. Thus it is not surprising that Urey went to work in a war-related industry, and that his brother Clarence served overseas in the army.

The church itself took an interesting stance. Brethren leadership advocated a "constructive patriotism" among its members, and encouraged draft-age Brethren men to consult their own consciences in determining a sacrifice "commensurate with the sacrifices made by those not exempted from military service."[98] Historians have interpreted this position as part of the church's ongoing process of Americanization. The church was taking steps to bring itself in line with mainstream Protestant denominations, although it still maintained its Anabaptist and Pietist interpretation of Christianity. Like other ethnic German churches that had already started down the path toward assimilation, the Brethren used the war and the emergent superpatriotism as an excuse to ignore conservative objections to abandoning the ethnic ways perceived by mainstream Christians as backward. They sought to present themselves as "100% American."[99]

If Urey needed any guidance on how he, as a Brethren man, should view the rise of tyranny in Europe and its threat to democracy, he might have gotten it from an essay that his father wrote sometime before his death. The essay, titled "Why Our Boys Should Be Patriotic," still survives in the possession of the family today. In it, Samuel Urey left no doubt about his own position. In addition to being a Brethren minister, Samuel Urey was a patriot who believed with great conviction that democracy could not survive without defense. Born just after the end of the Civil War, Samuel had developed a strong bias against Reconstruction-era Southern Democrats, and he felt sure that there were still more fights to come to preserve the Union. It would need patriots.

"Our imagination pictures before us such boys of the past," he wrote. "The noble Washington. The brave Adams and Hancock. The shrewd Hamilton and Jefferson. These are but a few bright names among a host of others." He also felt that his own boys would have a role to play in the future of the country: "Boys, the great work is not yet completed. We have many forts to hold and many assaults to resist. The anarchist is found wherever we may go. Impolitic leaders who seek nothing but the gratification of self are firing cannon against the very strongholds of our personal happiness, and social liberties, while foreign nations are lurking round us waiting but for an opportunity to smite our freedom."[100] The nation Samuel imagined needed patriots, and these patriots—his boys included—would have to be willing to fight for their freedom.

USEFUL WARTIME SERVICE

Having disassociated himself from the church already while in his "zone of invisibility," and having grown accustomed to fitting in with his non-Brethren peers, Urey decided to fight "on the chemical side," as Bray suggested. He later admitted that he felt some pressure to join his fraternity brothers in military service and to take part in the excitement of the war effort; he also felt torn between his impulse to prevent "Kaiser Bill and the whole German theory of government" from taking over the world, and the "early training of the pacifist character that I had received as a child."[101]

It was not difficult for Urey to find work in the industry during the war. Companies were actively recruiting American chemists to work in their wartime facilities; according to one who worked alongside Urey at the Barrett Chemical Company, Barrett's recruiters scoured the country looking for "anyone who was warm and breathing" and not already headed to the army.[102] Barrett had recruited Urey's freshman chemistry professor, Fred H. "Dusty" Rhodes, to head its research division, and Rhodes recruited Urey to follow him.[103] Alpha Delta Alpha listed the departures and military placements of its members in 1917/18, reporting that "Harold Urey received a good offer from the Barrett Coal Tar Products Company of Philadelphia, which he decided to accept. A. D. A. very much regretted to loose [*sic*] his company but was very glad to hear of his good fortune."[104] With the encouragement of his two professors, Urey went to work in Barrett's Frankford plant in Philadelphia, where he helped prepare chemical materials for munitions production, including toluene for trinitrotoluene (TNT).[105] In Urey's account, this was the moment at which he ceased to be a biologist and became a chemist.

Working in the chemical industry removed Urey from draft eligibility. American chemists as a professional group had spent a great deal of effort in print and lobbying the Wilson administration and the army to treat chemical work as an alternative form of war service. In addition to establishing the CWS within the army, chemists had sought exemption from conscription (or reassignment once in the camps) by promoting work within the chemical industries as a form of enlistment. This position is illustrated in a 1917 front-page editorial from the *Chemical Engineer*, titled "The American Chemist Must Enlist." Admitting that among Americans "the desire to do one's 'bit' is universal," the journal insisted that "the development of our national stamina and power is not a question of the existence of [America's resources], but rather one of their intelligent application."[106] The chemical community was such a resource, and the chemist "must be ready to give [his wholehearted efforts] without hesitation."[107] But the chemist should not by any "mistaken sense of duty" present himself at a nearby recruiting station; he should instead present himself to the national census of chemists, administered by the US Bureau of Mines and the American Chemical Society: "[By] all means let him send his name at once to be enrolled among those upon whom the government can call, and then let him be ready to submit his talents and training to the direction of our leaders at a moment's notice. It is in this way, and in this way alone, that he can be sure he is serving his flag to the best of his ability."[108]

The patriotic service that the American chemists constructed also came with a healthy dose of anti-German sentiment. Despite the oft-reported international character of science, chemists were not immune to the Americanism that ran rampant during the war. Just as the American public policed itself for un-American sentiment, so too did the chemists. Whether or not Urey had found superpatriotism attractive, he certainly would have gotten a strong dose of it in the chemical industry. In April 1918, for example, the distinguished Chemists' Club of New York announced that it intended to "measure up to the national standard of straight-out Americanism." Noting that the German element of the club had already, for the most part, quietly absented itself from the club premises, the club nonetheless felt compelled to proclaim that it sought "no companionship nor association with those whose allegiance lies with that country whose ruthless ambition has plunged the human race into a world war." The club henceforth forbade German nationals from attending, forbade the employment of German waiters in the dining room, and warned its remaining members and staff to beware of naturalized citizens who masqueraded behind their papers. "Hunt these down with every agency the country furnishes and with all celerity forbid them the doors of the Club," they wrote. "We are at the parting of the ways. If the Chemists' Club is an American Institution—make it truly such."[109]

It is perhaps fair to speculate that the Barrett Chemical Company—producing high explosives for the army—required at least as much vigilance against German sabotage as the Chemists' Club. Once Urey began working in Philadelphia, he likely had to be guarded about his life history. The work he did at Barrett was fairly basic industrial chemistry, but it was vital to the war effort.[110] A more interesting chemistry was at work, though; the war crystalized a chemist—and an unmistakably American one at that—from the persona Urey had developed in Montana.

CHAPTER TWO

From Industrial Chemistry to Copenhagen

There was little to hold Urey in Philadelphia once the war ended. He had not put down roots there, and he found little passion in his work for the Barrett Chemical Company, or in the idea of continuing his career in industrial chemistry. He stayed on as a research chemist in Barrett's Frankford plant for only one year before deciding to return to the University of Montana—the place where he had shaped his scientific persona and, in return, found a sense of belonging. Urey at first considered joining the faculty as a biology instructor. The departure of his former mentor, Archie Bray, had opened up a position within the biology department. He reached out to Morton Elrod, still department head, who felt that the young man would be a suitable replacement for at least some of Bray's duties, running the laboratory for the department's introductory courses while Mary Elrod picked up Bray's teaching duties.[1]

But the war had raised the prestige of American chemistry. University chemistry programs and their faculties had provided their services to the state. Public universities, already accustomed to serving local commercial interests, were particularly adept at mobilizing scientists for military service. Henry P. Talbot, professor of chemistry and chemical engineering at the Massachusetts Institute of Technology, asserted in the *Atlantic Monthly* in 1918 that the war was "preeminently 'a chemists' war,'" with chemists contributing not only to the research and development of gases and explosives, but also to food conservation and dye production at home.[2] In September 1918, reporting on the fourth annual National Exposition of Chemical Industries, the *New York Times* reported that American chemists were not only winning the war, they were also surpassing Germany in the production of quality dyes and other domestic products.[3] Universities hoped to keep up the increased chemical activities afterward, and built up their departments accordingly. The University of Montana was in need of chemistry instructors. Elrod noted to the university president that Urey had expressed a desire to "devote his time as a student to Chemistry and Physics."[4] So it was not as a replacement for Bray but as an addition to the chemistry faculty that Urey returned to Montana.

By Urey's account, he had little chance to pursue physical chemistry during his undergraduate training in Montana. However, he certainly was aware already of some of the exciting developments taking place in the field during these years. His primary exposure came from his chemistry professor, Richard Henry Jesse Jr. (also the university's dean of men). Jesse mainly taught his students analytical chemistry—what Urey described as "the regular, old-fashioned descriptive chemistry with not too much attention to physical chemistry"—but he also taught his students about Bohr's atomic model (introduced as recently as 1913). Urey "didn't understand it at all, not a bit of understanding whatever at that stage"[5]; nonetheless, he at least knew about the advances being made in Europe.

It is possible that Urey was more influenced by Jesse than he remembered, and Jesse was certainly more aware of physical chemistry than Urey realized at the time. Urey in fact recalled incorrectly that Jesse had done his doctoral work at the University of Illinois; between 1909 and 1912 he was a member of that university's faculty, not a student. Prior to this he had been one of Harvard University's rising stars. As a graduate student in Harvard's chemical laboratories, Jesse worked under the direction of Gregory P. Baxter, an American authority on atomic weights.[6] With a grant from the Carnegie Institution, and employing Baxter's "grand classical Harvard atomic weight technique," Jesse had determined a new atomic weight for chromium. Only after completing this work, and after spending a brief time as one of Harvard's teaching fellows, did he leave the East Coast.[7]

While at Harvard, Jesse was very near one of America's new centers of physical chemistry: Arthur A. Noyes's Research Laboratory for Physical Chemistry at the Massachusetts Institute of Technology (MIT). Here, throughout Jesse's graduate career, Gilbert N. Lewis, who would become

Urey's graduate adviser at Berkeley, was employed and was acting head of the laboratory during 1907 and 1908.[8] In 1910, Jesse published a paper with Lewis's Harvard graduate adviser, Theodore W. Richards.[9] Even though Jesse was not a member of Noyes's research team, he would have been well aware of the great interest surrounding the lab. He may have shared some of this excitement with the young Urey.

Urey's return to Montana was brief. After two years of teaching chemistry at his alma mater, he headed west to California. Precisely what made him decide to trade Missoula for Berkeley in 1921 is unclear, at least in his account of events. In his interview with Zuckerman, he suggested that the move to California was not entirely premeditated, but rather "was sort of a blundering decision, I would say, not well thought out."[10] In reality, Jesse likely nudged Urey in the direction of Lewis's California laboratory; in at least one account of Urey's life, it was Jesse who wrote to Lewis recommending Urey as a graduate student, helping him to secure the fellowship that allowed him to attend Berkeley.[11] Also significant was the fact that the entire Urey clan seemed to be moving west. The family had abandoned the Big Timber farm and headed to Idaho, where they would remain only briefly before Cora and Alva made their way to the West Coast. Martha, meanwhile, had just finished an undergraduate degree from a Brethren college in McPherson, Kansas; she would eventually become the dean of women at La Verne College—the Brethren college Eshelman had founded in Lordsburg. Clarence also made his way west to Oregon after returning from World War I.[12]

LEWIS'S LAB

> On both the theoretical and experimental side, thermodynamics first came west of Dodge City when G. N. Lewis went to Berkeley in 1912. Before 1912 the men west of Dodge City were busy defending their cows, their riparian rights, and their women; and they slept with their six guns.
>
> DON M. YOST, REVIEW OF *ELEMENTS OF THERMODYNAMICS*

When Urey arrived at the University of California, Berkeley, in 1921, he found a physical chemistry program that was quickly developing a reputation as one of the world's best. Don Yost's 1967 assessment of the state of science at Berkeley prior to Lewis's arrival is more myth than reality;

it did not mark a transformation of the American West from frontier to scientific center. Still, great transformations took place in the 1910s and '20s. Just as World War I transformed Urey from a biologist with chemical interests into a working chemist, it also accelerated trends that at the turn of the century began transforming the University of California from a little-known intellectual outpost into a contender among America's top research universities—especially in the area of physical chemistry.

Lewis was one of the most highly decorated chemists of the war. At the same time Urey was heading off to the Barrett Chemical Company, Lewis enlisted in the army as a major in the Chemical Warfare Service and was sent to France. Here he took up the position of director of the CWS laboratory in Paris and was later appointed chief of the service's defense division. His major accomplishment was establishing the American Expeditionary Forces Gas Defense School—a program designed to train "gas officers" that could help keep units alive during gas attacks. Gas casualties had accounted for the greatest number of troop losses, but thanks to the training program they soon contributed a relatively small percentage. This success in curtailing the loss of American troops, along with work on increasing the efficacy of American mustard gas attacks, earned Lewis a promotion to lieutenant-colonel and chief of the training division of the CWS.[13]

Germany's loss of the war led to a breakdown of the internationalism that had previously characterized European physical and chemical research, isolating German scientists from their colleagues and collaborators; but in the Allied countries, chemists were treated as heroes and introduced to new levels of prestige and resources.[14] Lewis and his Berkeley colleagues, who had succeeded in proving that they could "outgas the Germans[,] . . . enjoyed something of the reputation at the Armistice that atomic physicists did on V-J Day."[15] Lewis returned from the war to become "the dominant force for building up [Berkeley's] research capacity in physical science."[16]

After the war ended, California was able to maintain its heightened level of funding and research activity. According to the historians John Heilbron and Robert Seidel, nowhere in the country "were the consequences of the mix [of the interests of science, industry, and government during the war] more enduring and efficacious than in California."[17] Lewis took advantage of this situation. He had already brought dramatic changes to Berkeley's chemistry department even before the European conflict began. He had come to Berkeley from MIT in 1912 with the promise of complete control of the College of Chemistry; a brand-new, state-of-the-art laboratory; equipment funds; a salary of $5,000 (the highest salary within the physics department was only $4,000); a 50 percent budget increase to cover hiring new instructors; and a staff consisting of a mechanic, a glassblower, a bookkeeper, and an administrative assistant.[18]

Some of Lewis's East Coast colleagues predicted that his move to Berkeley would cut him off from America's centers of scientific activity, and would thus become a form of intellectual exile. Other scientists who had moved to California before Lewis had found little in the surrounding community to stimulate their work.[19] But Lewis had a tendency to be reserved and distant to begin with. He could be difficult to get along with and often put little effort into making friends. A certain amount of estrangement from his East Coast colleagues may have appealed to him. As one historian described it, "Lewis was uncomfortable in the larger scientific world and built his chemistry department at Berkeley as a support system for himself, staffing it almost exclusively with scientists who had trained there."[20]

What a support system it was. He brought with him from the East William C. Bray (no relation to Archibald Bray), Merle Randall, Richard C. Tolman, and in 1913 Joel H. Hildebrand and G. Ernest Gibson.[21] Lewis saw to it that his coterie enjoyed many of the same benefits as he. The increase in funds and intellectual manpower, not to mention the freedom from a heavy teaching load, quickly affected the activities of the department. In the 1912/13 academic year, four of the twelve members of the chemistry faculty produced eighteen papers, a significant increase from the five papers published by the faculty during the biennium, 1910–12.[22] Seven of these new papers were Lewis's. The department also became more attractive to graduate students during this time, and Lewis would use this to his advantage in sculpting his empire from within while severely limiting outside influence. By the time Urey arrived at Berkeley, Lewis had hired seven additional junior faculty members, all of them department graduates advised by himself, Gibson, Tolman, or Hildebrand.[23] Urey came to believe that if he distinguished himself in the field, he would also be invited to join Lewis's group permanently.

BREATHING IN THE BERKELEY AIR

When Joel Hildebrand wrote in 1963 of "the air Harold C. Urey breathed in Berkeley," he was speaking both literally and metaphorically. Gilbert Lewis's cigar habit was well known—he smoked "daily eighteen five-cent, Philippine cigars," the smell of which permeated every one of the College of Chemistry's laboratories and seminar rooms.[24] His philosophy toward chemical research and education was just as pervasive. This philosophy was forged through his experiences as a graduate student at Harvard and in Germany, and as a member and temporary director of Noyes's lab at MIT. These experiences led him to take a relatively free approach. Rather than attempt to control the college by fiat, he instead concentrated on building up talent and fostering an environment in which both students and faculty would feel encouraged to pursue new and interesting chemical questions.

Lewis had been a student of Theodore W. Richards at Harvard and had not liked Richards's close supervision. Richards held the opinion that "assistants who are not carefully superintended may be worse than none, for one has to discover in their work not only the laws of nature, but also the assistant's insidious if well meant mistakes." He was no more likely to give free rein to even his most brilliant students: "The less brilliant ones often fail to understand the force of one's suggestions, and the more brilliant ones often strike out on blind paths of their own if not carefully watched."[25] By contrast, Lewis avoided directly advising students and took a hands-off approach when directing their research.

Given Lewis's own predilection for solitude, it is unsurprising that he required little coursework of his graduate students. He "[allowed] the graduate student the greatest possible latitude . . . [and helped them acquire] initiative, morale, and a fine spirit of cooperation among themselves and the faculty."[26] Rather than being forced to sit through lecture courses, students were encouraged to find the information they needed in books. Seminars were offered on special topics, and these focused almost entirely on "hot" topics on which an instructor was preparing to publish.[27] Lewis made students participants in the cooperative work of the laboratory, as he had been in Noyes's MIT laboratory. Urey remembered, "It was immediately obvious that grades per se were not important things in this department. . . . I have never known what my grades were in any course that I took because it was unimportant."[28]

Graduate students in the department were in fact taught not to think of themselves as students, as the program Lewis had devised recognized few distinctions between his graduate students and his faculty. Rather, as Urey recalled, the size and structure of the program encouraged them to interact as colleagues: "There were only some 30 or 35 graduate students in the Department of Chemistry at Berkeley at that time. We were sufficiently few in number, and the professors were sufficiently few in number that to a large extent all of us knew everybody else—professors and students alike. I argued principles of physical chemistry with my fellow students and with my professors almost daily."[29] Hildebrand similarly remembered that in these early years, as the faculty was still happily adjusting to their reduced teaching schedules and increased research budgets, the amount of casual discussion between professors and students was remarkable: "The members of the department became like the Athenians who, according to the Apostle Paul, 'spent their time in nothing else, but either to tell or to hear some new thing.' Anyone who thought he had a bright idea rushed to try it out on a colleague. Groups of two or more could be seen every day in offices, before blackboards or even in the corridors, arguing vehemently about these 'brain storms.'"[30] This attitude toward free discussion of ideas and research became a hallmark of Urey's own approach to science and collegiality.

Nowhere was Lewis's egalitarian approach more evident than in the college's Friday evening seminar. "All of the people in the department attended—organic, physical and all—and none of us would have missed this seminar for anything else," Urey said. "Professors sat around a table in the middle of the room, and the students on a raised platform around the outside. Professor Lewis smoked that heavy cigar throughout the whole discussion."[31] At this seminar, anyone might be asked to discuss their research. Hildebrand recalled, "Lewis would say, 'Mr. Blank, will you tell us about your work?' Mr. Blank might be a member of the faculty or a graduate student; his research might be concerned with physical, inorganic, organic, analytical, or applied chemistry. No one could claim, as was often the case in Germany at the turn of the century, 'Das ist aber *mein* Gebiet.'"[32] A seminar participant might also be asked to report on new developments outside of Berkeley, as "any hot problem was fair game."[33] William Jolly described the seminars as typically consisting of two talks: the first a review by a graduate student of a published paper, and the second a presentation of original research being prepared for publication by a faculty member, advanced graduate student, or a postdoctoral student.[34] These weekly seminars, which constituted the extent of what might generously be called Lewis's "teaching," helped him keep all his chemists discussing and arguing the finer points of the latest work among themselves, and also allowed him to subtly hone the research ideas of the faculty and students and "thrash out" any inconsistencies.[35] Although Lewis was a problematic adviser for Urey and the two would eventually fall out, the younger chemist nonetheless adopted much of Lewis's philosophy toward running a research program and department—especially the Berkeley-style weekly seminar and the preference for an intimate cohort.

One of the most appealing qualities of the college after the war was the incredible freedom and resources Lewis bestowed on graduate research. Not only were the graduate students able to choose their research advisers on their own—and to change them whenever they pleased—they were also given "the run of the store rooms and laboratory facilities."[36] This was no small thing. By 1918, money from a state bond issue had built a state-of-the-art research facility that chemists from all corners of the world would envy: Gilman Hall.[37]

Upon Gilman Hall's completion, the *Journal of Industrial and Engineering Chemistry* ran a seven-page spread, complete with floor plans and photographs, of what was surely one of the world's best-equipped university chemistry facilities.[38] The building was industrial in character, consisting of two main floors and a furnished attic, along with a basement, subbasement, and sub-sub-basement. The walls, floors, and roof of the building were all built of reinforced concrete. The rooms—labs, offices, and classrooms—were each assigned a specific purpose in the building's floor plan, but were also built in such a way that significant changes could be made at any time. All the building's piping and electrical conduits, for example, were exposed on the ceilings to facilitate any changes that a space might require, and removable wooden panels had been placed in the walls separating the rooms to facilitate further reappropriations of space.

Aside from being accessible, the piping and wiring systems were also versatile. Each lab was supplied with gas, low-pressure air, suction, oxygen, and distilled water. A researcher could easily outfit his lab with high-pressure air, vacuum, steam at various pressures, liquid ammonia, crude fuel oil, and electricity of varying voltages in AC or DC current (the generators for which could be controlled from any room). Well-appointed faculty offices were attached to each lab. In addition to providing ideal laboratory space, the building also boasted large and well-outfitted shops, including a variety of craftsmen's shops, an instrument shop, a glassblower's shop, and a shop within which students could machine their own tools and instruments.[39]

Urey thrived in the Berkeley atmosphere and enjoyed access to the college's experimental resources, but he also felt his dissertation project would have been more successful had he been more closely advised: "I never quite knew whom I was working with; whether I was working with Lewis or whether I was working with Olsen, and I don't think Olsen ever knew exactly. . . . Lewis didn't watch very much what I was doing, and nor did anyone else."[40] At Lewis's suggestion, Urey had taken on a dissertation project that was a thermodynamic study of the conductivity of molecular cesium vapor.[41] Urey's work was difficult and butted up against the limits of quantum mathematical tools that were still being developed. While physicists were approaching this work theoretically, Urey was approaching the problem empirically, using spectroscopic data.[42] But while he ultimately was disappointed by his doctoral dissertation, he did feel that the project and his two years as a graduate student in the Berkeley environment gave him his "real start as a research scientist."[43] He also may have underestimated the success of these early experiments, as they did provide the foundation for his later award-winning work.

QUANTUM CHEMISTRY FROM CALIFORNIA TO COPENHAGEN

Urey might not have been proud of his dissertation, but it did provide him with two publications—one on the heat capacity of gases and the other on their ionization.[44] Furthermore, Urey's interest in these two subjects had been piqued, and he had become acquainted with the theoretical properties of atoms and molecules. He had also, thanks to his laboratory experience, become quite adept at the technique of molecular spectroscopy. But most important, Urey's work at Berkeley exposed him to the newly emerging field of quantum chemistry—a field that American chemists would come to dominate in the 1920s and '30s.[45] Following on the heels of chemists like Richards, Noyes, and Lewis—the generation of American chemists who had studied physical chemistry under Wilhelm Ostwald and Walther Nernst in Germany and who imported the field to the United States—a new generation was rising that would usurp some of Germany's authority in pure chemical research.

From Niels Bohr's 1913 paper on the hydrogen atom into the 1920s, the quantum physicists of Europe seemed intent on laying out the theoretical principles within which the facts and laws of chemistry could be fit. Their goal was to use their physical theories and mathematical formulae to "create a mathematical and theoretical chemistry."[46] However, as the historian Mary Jo Nye points out, very few of these physicists had any meaningful understanding of chemistry. German chemists contributed very little to the emergence of quantum chemistry. Instead, a roster of young American chemists and physicists, including Urey, made the most significant contributions to this new field.[47]

Developing the field required the collaboration of physics and chemistry. In this respect, Urey was fortunate to have studied at Berkeley when he did. The physicist Raymond T. Birge had arrived at the university during World War I and, when the chemists returned, had worked with Lewis to forge new lines of communication between their respective departments. In spring 1920, the year before Urey arrived, when Birge and two of his fellow physicists offered a course on "Radiation and Atomic Structure," several members of the chemistry department attended. In the following year, Birge's course became a full-year course with a heftier title: "Radiation and Atomic Structure: A Discussion of Recent Work in the Fields of Electric Discharge through Gases, Spectroscopy, X-rays, and Magneto-optics, Bearing upon the General Problem of Atomic Structure." This course attracted the attention of chemistry staff and graduate students alike, and "presently every graduate student majoring in physical chemistry was sent to take the course for credit."[48] Also in the physics department, William H. Williams began offering his graduate seminar in theoretical physics—a course that remained popular among physics and chemistry graduate students up through J. Robert Oppenheimer's tenure at Berkeley.[49]

Urey found Birge to be "a very inspiring professor." He became one of the three greatest influences on Urey's scientific development at Berkeley—the other two being Lewis (who held himself at a distance) and Hildebrand. As Urey said in an interview, Birge was "the one who interested me a great deal in physics by (running) seven hours of courses and talking to me about the Bohr atom at the time."[50] The two came to know each other well. Their discussions of physics were not limited to the classroom, but also took place on walks along the idyllic canyon trails in Muir Woods.[51] A letter of recommendation Birge wrote to Niels Bohr on Urey's behalf in 1924 gives some idea of Birge's impressions:

> Dr. Urey is a very unusual man. He started as a chemist, but has since shown remarkable interest and ability in mathematical physics. I should call him now a mathematical physical chemist. From my own knowledge—not as extensive as it might be—Prof. [Richard] C. Tolman is perhaps the only other man in this country who should be similarly classified. I mention this mainly because Prof. Tolman is perhaps the only one in this country who is really qualified to judge Dr. Urey's previous research work, and he has in print referred to it in very complimentary terms.[52]

Birge here indicates that Urey's work had already caught the attention of another California researcher, Caltech professor Richard C. Tolman. Tolman was one of the leading authorities on physical chemistry in America at that time. Along with Birge, Lewis, Richards, and Noyes, Tolman was teaching and advising the new generation of American physical chemists; while in Berkeley Birge and Lewis were shaping Urey into a quantum chemist, in nearby Pasadena Tolman was advising another future luminary (and Urey's later collaborator), Linus Pauling. Tolman credited Urey in the pages of the *Journal of the American Chemical Society* with advancing the development of the quantum equation for the entropy of a diatomic gas. In an eight-page article, Tolman dedicated nearly half a page to recognizing Urey for his application of experimentally derived spectral data to what had until that point been a highly theoretical question in physical chemistry.[53]

After providing brief descriptions of Urey's published work (the two papers that his dissertation comprised), Birge went on to describe Urey's initiative in bringing this work to completion without the benefit of external direction. Birge also noted that, by doing so, Urey had gone beyond even the understanding of his faculty advisers:

> To me the most significant point in this connection is the fact that Dr. Urey worked up this material with practically no assistance. I helped him as best I could as to the general facts regarding atomic structure, but I freely confess that a year ago my knowledge of the statistical side of these problems was lamentably poor. . . . I am just beginning to appreciate the importance of Dr. Urey's work. I have never had any question as to the brilliance of his intellect, but when he talked over his problems with me, I frankly was not familiar enough with the matter to give him any real help. I trust I will not be misunderstood when I add that there was no one else here at Berkeley who was familiar enough with this field to help him.[54]

Birge finished by likening Urey to two already well-known young pioneers of the quantum movement, John C. Slater and John Hasbrouck Van Vleck, both of whom received their PhDs at Harvard while Urey was at Berkeley. Birge predicted that these three young men would "bring America's ranking in mathematical physics up to the European level."[55] Indeed they would; all three would contribute to the evolving understanding of atomic structure, and two would win Nobel Prizes for their work.

THE COPENHAGEN SPIRIT

The young men of science who traveled to Copenhagen in the 1920s would turn the tables on their European colleagues. The Americans who spent time in Bohr's institute as students would return to the United States and develop a true quantum chemistry.[56] They would help move American physical science from the periphery to the center of the world scientific community. They would be of the last generation who felt obliged to travel to Europe for study. When émigré scientists began leaving Europe during the 1930s, Urey and his peers would play host to them in the departments they had either built or transformed, proud to at last be able to introduce the international component to the salon model they had brought back with them from Copenhagen, Berlin, and Göttingen. But when these young Americans first arrived in Europe, they had little to recommend them as revolutionary scientific minds.

Most of the early group that traveled to Copenhagen in the 1920s found that they were ill prepared for the intellectual work of Bohr's institute. Urey discovered that Birge had overestimated his grasp of mathematical physics. He wrote to Birge in the 1960s that "one of the things I learned in Copenhagen as a result of my association with [Hendrik A.] Kramers and Slater and [Wolfgang] Pauli and [Werner] Heisenberg was that I could not do theoretical work and keep up with these men."[57] He told Heilbron, "I did not have the mathematical equipment and the mathematical ability to be an effective theoretical person."[58] And in his unpublished autobiography, Urey wrote that "I tried to do some theoretical physics at Bohr's Institute. . . . Perhaps the most important thing that I learned was that these friends of mine at the Institute . . . would be able to do a much more thorough job in theoretical physics than I could possibly do, and that my strength in science would lie elsewhere."[59] Indeed, his contributions to quantum chemistry would be in the experimental realm.

Urey's uncertainty about his future in quantum chemistry, and his unease with the mathematics he was now expected to master, shows in his correspondence from this period. He delayed writing to Lewis from Copenhagen because he had nothing optimistic to say about his work there or the work he had left behind in California. In his first letter, Urey suggested that perhaps he should have stayed at Berkeley an additional year to continue the unsuccessful conductivity experiments with cesium vapor that had been part of his dissertation, or to take on a new experimental problem. Tolman and Birge had both praised his experimental work with cesium, but Urey could only see its deficiencies. He was convinced that Lewis was disappointed in the lack of results, and with Urey for his failure to design an experimental apparatus that would not interfere with the conductivity measurements. "I did not know what to say to you in regard to the experimental work," he wrote, "and I do not know now."[60] As for his new mentors in Copenhagen, Urey was relieved to report that Bohr would be leaving for England, telling Lewis, "As there are a great many things that I must read and study before I really grasp and understand the quantum theory, I feel that my time can be very well spent."[61] He confessed in a letter to Birge written a week earlier that he was finding Bohr's publications to be a difficult read, and noted, "These men have just grown up it seems with mathematics where we would have breakfast food."[62] Urey would not have his first substantive conversation with Bohr until almost four months later.[63]

Although Urey was butting up against his own mathematical and theoretical limitations, he nonetheless enjoyed his time in Copenhagen and felt personally transformed by his experience. "One of the most valuable things to me so far has been to see how people live here," Urey wrote to Lewis. "I believe that they know better how to live in many ways than we do."[64] Urey basked in "Der Kopenhagener Geist," the "spirit" embodied by Bohr's institute and its coterie. Bohr designed it to be "a place free from nationalistic emotion, which would revive and reaffirm scientific internationalism," an "international gathering place for physicists in the postwar period,"[65] and "an environment of vigorous intellectual engagement and affectionate esprit de corps."[66] Within its walls he had gathered a "modernist enclave."[67]

Like the other Americans who came to the institute during these early years, Urey was only a marginal participant in the international collaboration underway there.[68] But he participated as best he could. Unlike some of his fellow Americans from East Coast universities, he likely felt comfortable right away with the informality of the institute, its lack of classrooms, papers, textbooks, and laboratory exercises, and its insistence that physics was a "series of conversations."[69] He later jokingly told colleagues that he learned physics in the cafés of Copenhagen while dining with Bohr's assistant, the Dutch theorist Hendrik A. Kramers, who was his primary contact at the institute.[70] Kramers—described by Shannon Davies as "the prototype of the cultured pipe-puffing European physicist"—took an interest in Urey, despite his mathematical shortcomings, and the two met for lunch on an almost daily basis.[71]

Urey reported in a New Year's letter to Lewis that Kramers had become his main instructor, and Urey seems to have been quite taken with the Dutch physicist. He described to Lewis in detail all the many languages that Kramers spoke, his prowess as a physicist and mathematician, and his proficiency on the piano and cello. Rather jealously, Urey remarked that "he is primarily a physicist and always keeps the physics of a problem in mind and his skill and knowledge with mathematics makes it a most powerful tool with which to attack his physics. Professor Bohr remarked one day that [Kramers] remembered all the mathematics he ever saw."[72] Still, Urey noted, Kramers had "a pleasing personality and is very modest about all his many accomplishments."[73] Urey also reported that he had eaten lunch and had afternoon tea with Kramers almost every day for three months, had met him in his home many times, and had "talked physics a great deal."[74] Under Kramers's instruction, Urey acquired some advanced mathematics and attempted to do theoretical work on an orbital problem. He developed a working understanding of quantum physics and found that he could read the literature being produced in Copenhagen. He reported to Birge in early December that "so far I have been unable to get anything done but have progressed so far that I can read Bohr's and Kramers' papers with a great deal of pleasure. That at least is something."[75] Soon he was even able to write and publish one theoretical article with Kramers's editorial help.[76]

In addition to Kramers, Urey also met several of Bohr's closest collaborators. He forged a lifelong friendship with the Hungarian physical chemist George de Hevesy, who was in Copenhagen working on the properties of hafnium. He met the German theoretical physicists Werner Heisenberg and Wolfgang Pauli when they visited the institute. And after his time at the institute was up, he traveled around Europe introducing himself along the way to the German physicists Albert Einstein and James Franck, and the Swedish chemists Svante Arrhenius and Theodor Svedberg.[77]

Urey became incredibly fond of the city of Copenhagen, which seemed to represent something other than the "old Europe" where Lewis and Irving Langmuir had studied, and yet was also different from the industrial American cities he had come to know: "Copenhagen, during the 20s, was a city of bicycles, and I bought myself a bicycle which I used to travel . . . [to] the Institute daily, but in addition to that used it on the weekends to look over the whole city of Copenhagen. I marveled at the wonders of this city. There were poor sections of the city, but there were no ghettos."[78]

His hosts during his stay were the Danish historian Aage Friis and his family, and he lived in their absent son's room. Aage Friis was at that time engaged in trying to rebuild the international character of the European historical community, which in the early 1920s was still suffering from the effects of World War I.[79] The Friis family accepted Urey as one of their own, and although they spoke little English and Urey spoke no Danish and only limited German, he came to feel at home with them; through them he cultivated an appreciation for Danish culture and the European welfare state.[80] Urey's letters to Birge described the Friis household, which was no doubt very different from the small and spartan houses in which he spent his childhood, and also alluded to their hikes together back in Berkeley:

> This home is most beautifully decorated. They really have hundreds of pictures on the walls, all copies of very good art, beautifully framed. Each member of the family has a room which is his or her own and the decorations affect the taste of the individual. Professor Friis' study (in which I am writing) has all four walls mainly covered with books mostly unbound and the remaining portions literally covered with pictures of historical figures or historical places. He has certainly made himself a scholar in the history of Central Europe and the Scandinavian countries. . . . The downstairs belongs to Mrs. Friis and is decorated so abundantly that I could not begin to describe it all, with pictures and furniture of very beautiful kinds. The yard reminds me of Berkeley for its abundance of flowers. One could just settle down in a home like this and never wish for an automobile nor a movie nor any of the other numerous diversions of America except that of hiking and the Danes like hiking as well as we do.[81]

Urey may have recognized in the Danes something similar to the communal way of life he had enjoyed as a boy, and his bicycling no doubt would have brought back memories of cycling between Kendallville and the Cedar Lake farm. In some ways the Danish lifestyle he coveted was a sophisticated and secular version of Dunker life. He felt an immediate affinity for it.

In addition to their poor mathematical training, the young Americans in Copenhagen were faced with the relatively low status of American science in Europe. The physicist I. I. Rabi recalled that American physics was held in contempt during the late 1920s when he was in Europe. In Hamburg in 1927, Rabi sought the latest issue of the American journal the *Physical Review* in the library, only to find that it was not held in high esteem; the journal was ordered and shelved once all the year's issues had been published. Rabi concluded that Europeans did not feel it was important to have constant contact with the latest American discoveries in physics, since most major discoveries were still being made in Europe.[82] Having been trained amid the triumphalist science boosterism of the post–World War I period in America, this must have come as a shock to the young Americans.

The result of Rabi's encounter with Europe and its anti-American chauvinism was to "feel the greatness of America": "The Germans' misunderstanding of the United States was so great that I was known as a chauvinist because I would argue with them all the time. There were plenty of things in the United States I didn't like—plenty of things. But not what they talked about. They didn't like the things about us that were good. After all, we had an honest-to-goodness democratic system—you could live in it."[83] Rabi, having been raised and educated in New York, was already more cosmopolitan than Urey, and this no doubt left him less deferential or in awe of his new colleagues. Urey, who practically adopted a Danish view of the world (with a hint of Dutch, perhaps, via Kramers), had a very different experience. But Urey also detected some bias against American science. In his first conversation with Bohr, the physicist told him that the scientific "lead" Europe possessed over America would continue for many years "if Europe could settle its financial and international difficulties."[84] As Urey reported to Lewis, "He said he found so much enthusiasm and wealth in America but that science lacked the prestige which it has here."[85] Rather than contradict Bohr's impression of American science, Urey agreed with him, albeit with optimism for the future of science in America: "I have noted myself the respect paid to science and to learning in general here and have found quite a contrast with the attitude in America. However I believe that that prestige is coming and that America will acquire more standing in the theoretical fields as well as in the experimental work [for] which Europe [is] recognized at this present time."[86] Having invested so heavily in his own Americanness, colored as it was by his newfound love of Europe, this attitude seems completely appropriate.

Urey's interactions with Friis, Bohr, Kramers, and de Hevesy would help him redefine his own views of life and "civilization." As far as he could determine, aside from the absence of slums, "the primary difference in the economic standard between Danish families and American families was in our ownership of automobiles."[87] He later wrote to Kramers from Baltimore that he missed the lifestyle he had known in Copenhagen, and expressed his disappointment that he could no longer travel by bicycle but would have "to follow the crowd and buy an automobile."[88] He lamented that "people enjoy themselves here in such different ways."[89]

This love of Denmark stayed with Urey for the rest of his life. His experience in the small country, while it made him no less patriotic, did lead him to question whether at least some of the values espoused by his fellow Americans were not misplaced. And in cases in which he did believe Americans to be a bit backward or wrongheaded, it may have been easier to appeal to the Danish way of life than to the Dunker religion. But such critiques would not emerge in full force until the Cold War. For the time being, as Urey's star rose at home, his all-Americanness was a defining characteristic of his emerging public persona.

CHAPTER THREE

From Novice in Europe to Expert in America

> I had quite a bit of trouble getting started. You see, I thought I was going to be a theoretical chemist when I went to Copenhagen, and I just concluded I couldn't do this thing. So I had to get back and start a completely new line of work. And it took quite a while before I got started at it.
>
> HAROLD C. UREY, IN AN INTERVIEW WITH JOHN HEILBRON

Urey was anxious as he prepared to leave Copenhagen in the spring of 1924. He wanted to return to Berkeley, to a secure position within which he could begin a full-time career in science. Throughout his education, he had made his home first at the University of Montana and then at Berkeley. He did not consider himself to have a home outside these institutions. As he explained to one historian:

> I was always on the move. While I was at the University of Montana, my parents moved from Montana to Idaho. I had never lived in Idaho, so I could not call that my home. My home was the university. I went to Philadelphia. That was my home. I went back to Montana and that was my home again. I went to Berkeley and they had tuition for out of state students. I said I came from Montana, but my home was not in Montana. My home was in California now.[1]

Urey had soaked in the atmosphere of Bohr's institute, and now he wanted to return home as one of Lewis's coterie. Lewis responded to his request to return with mild encouragement, noting that he would like very much to have him back in California, but he suggested that Copenhagen may have made him more of a physicist than a chemist, and so perhaps the physics department would rather have him as a lecturer. This suggestion did not please Urey, who felt that his knowledge of physics was "entirely too localized to fit me for a physics department." In the years to come, Lewis and his group would become increasingly interested in the new quantum mechanics, but for the time being it seems that Lewis was uncomfortable with the physicists for privileging mathematics over physical models of the atom. Lewis admitted as much to Urey only a few years later; in a critique of Erwin Schrödinger's work, he noted that "the history of physics seems to show that continuous progress can not be made in such [mathematical] ways alone, but that some model or picture will again be necessary in order to furnish some sort of visualization of abstruse laws and to suggest new experimental tests."[2]

Urey was also unhappy with Lewis's suggestion that he apply for a fellowship from the National Research Council. "This idea does not please me at all for several reasons," he told Lewis.

> Such fellowships require one to do some complete research within the year. The necessity of producing something places me under a nervous tension that is quite unpleasant. . . . In the second place, tho I am not particularly old, nevertheless I am far enough to wish to get started someplace and I feel that a National Research Fellowship is only standing still for a year. For a man at the age of twenty three, say, this is not true but for me it is. For these reasons I would prefer to get a position in a smaller university or college if necessary and so be able to take root and grow fast for a few years at least.[3]

Urey's discouragement was short lived. Unexpectedly, Bohr invited Urey to spend another year at the institute. This additional year never materialized, but in the short term it gave Urey the courage to write to Lewis to ask why he was not being offered a position, and to begin making plans of his own.[4]

Whatever answer Lewis ultimately may have given for not offering Urey a position—if he gave any at all—is not on the record. Given Lewis's predilection for staffing his department with his favorite former students, Urey must have felt jilted. By July Urey had reconsidered the National Research Council fellowship and decided to apply. He would go to Harvard with John Slater, whom he had gotten to know while in Copenhagen, and whom he reported to Lewis was "a mighty fine chap."[5] Urey wrote to Edwin C. Kemble (adviser to Slater, Robert Mulliken, and Van Vleck at Harvard) to tell him that he was applying for a fellowship and wanted to spend a year under his direction working on a theoretical problem begun with Kramers.[6] Urey was offered the fellowship and did appear in Cambridge at the end of his European tour, but he spent less than a month there before leaving to accept a position Lewis secured for him at Johns Hopkins University.[7]

Urey later stated that he gave up on theoretical work while in Copenhagen. And, indeed, even before he left Copenhagen he seems to have decided, like Lewis, that much of quantum theory had "left physics entirely and [had] gone entirely or nearly so to the field of mathematics without any physical basis."[8] He felt reassured in this view by Einstein's rejection of the emerging Copenhagen interpretation of quantum mechanics. But Urey's letters show that even after he began at Hopkins, he was still attempting to do theory—with the hope that he could at least keep up with the American theorists. He had been happy to take up a position at Hopkins, but he regretted leaving Harvard before taking full advantage of "one of the most stimulating places I have been."[9] He seems to have believed that Kemble could make him Slater's theoretical equal. By mail, up through February 1926, Urey continued to pursue Kemble's advice on theoretical matters and sent him copies of papers he wanted to revise for publication. At the same time, he also sent these papers to Slater, Mulliken, and Bohr.

The process must have been demoralizing, as Kemble and Mulliken's responses primarily contained corrections of errors in Urey's math and Bohr put off responding for quite some time. On his end, Urey struggled to keep up with Kemble's theorizing, and also deferred quite often to Kemble's authority. Additionally, Urey struggled to answer the questions Kemble asked him, only to be told in a following letter that it had been a "simple" matter. In one letter, nearly every sentence Urey wrote is apologetic. Perhaps he felt as though his slow learning curve was trying Kemble's patience and that every correction the physicist made only opened the door for more difficulty: "I believe that I see your line of argument at last and alas think that there is an ambiguity in the whole thing yet. I am rather anxious about this for if you are right I am in for a hullova [*sic*] time. . . . Do I have this right now? Then it seems to me that everything you say follows."[10] Finally, in February 1926 Urey gave up. He sent one last letter to Kemble, which ended with the admission he had shifted gears: "Have been working on some experimental work for the last few weeks and find it very interesting after working on theoretical things for so long."[11]

At the same time that Urey was giving up on theoretical chemistry and getting back to experimental work, his life was changing in other ways. In summer 1925, Urey had traveled west to visit his mother in Seattle. On this same trip, he also decided to visit Kate Daum, a woman he had known as a student at the University of Montana. Kate introduced him to her younger sister, Frieda. For two weeks they went on treks through the Cascades, getting to know each other as they hiked. Frieda was working as a bacteriologist in a doctor's office—she understood science and held her own in conversation. Moreover, she was a fellow midwesterner, born and raised in Lawrence, Kansas. She was impressed by his intelligence and excited about his world travels. The two fell in love. For Urey, whose letters to Birge often expressed a fondness for hiking and a rejection of automobiles and American urban life, this must have been the ideal courtship in the open air of the Pacific Northwest. He insisted that they marry at once and that she accompany him back to Baltimore, which she refused to do, having just met him. He returned to Baltimore without her, but they continued their courtship by mail until, in June 1926, the two were married at her father's house in Lawrence.[12] Married life agreed with Urey and gave him the confidence to assert himself scientifically. In 1927 he and Frieda had their first child, a daughter they named Gertrude (but who became known as Elizabeth). Two years after this, a second daughter, Frieda, was born.

Even though Urey found it difficult to do original work in quantum theory, he ultimately discovered that he had a gift for teaching it. This was a task he at first took up out of necessity. At Hopkins he was dismayed to realize that chemistry graduate students were largely ignorant of atomic physics and of mechanics in general. In a 1929 roundtable discussion on "The Teaching of Atomic Structure to Physical Chemists," Urey reported that there was "no doubt that chemists entering our graduate schools do not know any mechanics worth mentioning and it is completely useless to talk about teaching wave or matrix mechanics to people who do not know any classical mechanics, no matter how desirable that would be."[13]

Likewise, their math skills were well below par: "It is certainly true that graduate students of chemistry have forgotten most of the mathematics that they ever knew and, what is more discouraging, they have in many cases acquired a fear of the subject."[14] This bothered Urey, who wanted "to see physical chemists have some of the fun in the revolutional [*sic*] developments" in quantum mechanics.[15]

As Urey saw it, there were only two possible roads for chemistry as a profession: either the chemists' training should be revamped to include greater amounts of theory and mathematics—the road of the "pure scientist"—or chemists should "admit that we and our science will take the same position relative to physicists and physics that engineers and their subjects now hold relative to the latter."[16] At Hopkins, Urey chose to create pure scientists. Here he instituted a required course on atomic physics within the chemistry department. This course had "indifferent success" in its first incarnation, but Urey was able to improve it by inserting six weeks of mechanics into the beginning of the course before ever even addressing atoms or molecules.[17]

Urey's new course was inspired by his time in Birge's classroom. He also borrowed elements from his experiences with Lewis and Kramers, adopting the seminar approach to graduate instruction. As he explained in the roundtable: "I have found in giving a lecture course that the lecturer works very hard, but the students do nothing at all. The lazy method of sitting back and letting the students give reports on papers and subjects is the quickest way of getting a group of students working on the subject and they learn far more."[18] Just as at Berkeley, the seminars were focused on recent research and hot topics. During the academic year 1927/28, Urey's seminar addressed the recent work of the German physicist James Franck on the effect of light in dissociating molecules. In the following year, the topic was the study of band spectra.[19]

Urey also attempted to make inroads into the Hopkins physics department. Although intended primarily for the benefit of his chemistry students and therefore emphasizing the "experimental side" of research on atomic structure, Urey's courses nonetheless attracted a number of physics graduate students. He also made overtures to the physics faculty. He attended the physics seminar on a regular basis and there befriended a few of Hopkins's physicists—including Robert W. Wood, Karl Herzfeld, and Frank Price. On the whole, however, Urey found the physics department to be "pretty old-fashioned" and under the "imperial" grip of Joseph Ames, who was at that time also a university provost and the head of the Physics Laboratory. Urey's impression was that Ames and his physicists were not interested in the new quantum mechanics, nor did they seek to understand it: "No one could say anything that Ames didn't like, and Ames was very much of a classical physicist. And Wood, of course, knew nothing about modern Physics. . . . It was pretty much of an old fuddy-duddy department in a certain way."[20] Elaborating on the deficiencies of Wood, whom he considered to be one of the department's most talented physicists, Urey later told Heilbron that Wood simply "never understood quantum physics at all": "His experimental ability and a correct instinct for what was interesting [were his great strengths.] . . . But Wood never understood it in modern terms."[21]

Urey may not have been impressed with these older physicists, but he nonetheless began to form his own coterie of young atomic scientists during his four years at Hopkins. The German-born theoretical physicist Maria Goeppert Mayer and her husband, the physical chemist Joseph Edward Mayer, arrived at Hopkins at the end of Urey's time there. Urey was incredibly impressed with the Mayers, who would eventually follow him to Columbia and the University of Chicago; all three scientists would end their careers at the University of California, San Diego.

Urey mostly spent his time with his fellow research associates, including F. Russell Bichowsky. Before arriving in Baltimore, Bichowsky had worked in the geophysical laboratory of the Carnegie Institution. He knew the scientists in Washington, DC, and was a regular attendee of the National Bureau of Standards' weekly seminar. Urey accompanied Bichowsky to these seminars, which soon became more important than the "fuddy-duddy" physics seminars in Baltimore. At the Bureau of Standards he met the geophysicist Merle Tuve, the x-ray crystallographer Ralph W. G. Wyckoff, and the physicists Arthur E. Ruark, Ferdinand Brickwedde, Otto Laporte, William Meggers, Samuel Allison, Paul Foote, and Fred Mohler. Many of these men—the physicists especially—would become part of Urey's core group of colleagues and collaborators over the next few decades.

At these seminars, the participants chose their own presentation topics. One of the hot topics was the work coming out of Copenhagen. Here Urey found that one of his great talents was in explaining the new publications in quantum mechanics. While Schrödinger's mathematics and Heisenberg's new wave mechanics admittedly went beyond his own abilities, he nonetheless gained a reputation within this circle as someone who could decipher what these developments meant to practicing physicists and chemists.

Urey's explanations of quantum mechanics would not remain limited to the classroom or the intimate seminar format. To reach a wider audience, Urey teamed up with Ruark, who was at that time a member of the Atomic Structure Section of the bureau. The product of Ruark and Urey's collaboration from 1926 to 1929 was the book *Atoms, Molecules and Quanta*. Urey at first intended to create "a text book in mimeographed form," that would follow the outline and structure of his course lectures. But the mimeograph soon became a nearly eight-hundred-page monograph. While this book contained no original work, it was one of the first significant attempts to explain the advances in quantum mechanics to English-speaking scientists who were not specialists in the field. The two men split the work according to their respective skills, "with Ruark taking somewhat the more mathematical side of the problem and [Urey] the more descriptive sides."[22] As with the courses Urey designed at Hopkins, this book began with the experiments in classical mechanics that led to quantum mechanics; only the last two hundred pages addressed the most recent contributions of Bohr, Schrödinger, and Heisenberg. Reviewers of the book noted that it went "a long way toward filling this very real need [for a summary of the new atomic physics]," that it was written in a style "comprehensible to the reader who has not been trained in advanced physics," and that it presented both the experimental and theoretical aspects of the subject in such a way that both chemists and physicists would find the book useful.[23]

In typical fashion, Lewis's advice and encouragement to Urey when he began this book project were ambivalent. "If you decide to write the book I shall be very glad to see it, and I am sure it would be useful," he wrote to Urey.

> But if you wish my advice regarding the writing of it, viewed from your own personal standpoint, I should be rather inclined to advise against it. Writing a book, with all the proofreading, correspondence with the publishers, etc., always takes a great deal more time and energy than one expects. . . . It may

> be very well for the person who is approaching the unproductive years, but for a young man I think that it takes more out of him than the results are ordinarily worth.

Lewis concluded his advice with the observation that Urey's state-of-the-art knowledge of quantum mechanics might become out-of-date by the time the book was published. He encouraged Urey to pursue his experimental work instead. "If a person has no experimental work under way there often come periods when the pure theorist seems to come to the end of his rope, and it is rather discouraging. Experimental work, on the other hand, tides a man over these periods and does not, I think, make him less productive on the theoretical side."[24]

Fortunately for Urey, he ignored Lewis's advice and published the book. He decided after the fact that Lewis had been right, that the undertaking was too ambitious for a young scientist, and he swore to Birge that he would never try anything like it again.[25] Nonetheless, it further cemented Urey's reputation as an American emissary of quantum mechanics. Although he had yet to make a lasting or original contribution to quantum chemistry by the time he left Hopkins, Urey had become one of the young leaders in the field. His math skills might have kept him from joining the ranks of Heisenberg or Schrödinger, but his few publications on the statistical thermodynamics of gases had marked him as one of the promising young American chemists who could apply theoretical work to traditionally chemical topics. Moreover, his ability to explain quantum mechanics to the uninitiated had moved Urey from the periphery of the Copenhagen circle to the center of the American physical science community (even if he did still defer to some of his more mathematically minded colleagues at home).

In the same year that the Ruark and Urey volume was published, Urey was made the founding editor of the *Journal of Chemical Physics*—a publication started by the American Institute of Physics with the purpose of providing a home for work not then being published by the American Chemical Society or the *Journal of Physical Chemistry*.[26] This editorship gave Urey—who had already managed to publish twenty papers or notes about atomic structure and experimental molecular band spectroscopy while at Hopkins—a key role at the intersection of the chemistry and physics communities, and in the creation of "chemical physics" and quantum chemistry as disciplinary fields. However, while his status may have risen high enough to be entrusted with the editorship of the new journal, he was not yet the scientist who would be regarded as the logical choice to head up the wartime effort to separate uranium's fissionable isotope.

Accounting for Urey's quick professional rise within American academia, Kevles claims that Urey's time in Copenhagen had "completed his transformation from an uncertain neophyte into a bantam cock of a physical chemist."[27] He was a productive researcher, to be sure. Urey's experience, however, attests more to the importance of the reputation he acquired simply from his association with Bohr's institute and its circle of European physicists. In this he was not alone. Rabi experienced something similar upon his return from Europe. When he started as a lecturer at Columbia in 1929, Rabi had accomplished little but "was the life of the place": "Students were flocking around, and I was in correspondence with and close to other physicists who were well known, and so on. I was in the mainstream."[28] Rabi credited his "mainstream" status with the fact that, "after the first year, even though I didn't publish anything, I was given a promotion to assistant professor. . . . After the second year, I was given a raise. The third year, I still hadn't published much of anything, and they wanted to make me an associate professor."[29] Urey felt similarly when, after accomplishing very little at Hopkins, he was offered a job at Columbia University in 1929.

Urey's assessment of his time at Hopkins was dim. Much of the work Urey performed while in Baltimore he would later refuse to cite, telling Ruark, for example, that he would not reference papers that were incorrect, even if they were his own.[30] However, Urey did manage to publish a handful of research papers during his Hopkins years dealing with the applications of molecular spectroscopy to chemistry, including one on the structure of the hydrogen molecule ion.[31] He also took the opportunity to collaborate with his two fellow research associates, Bichowsky and Francis O. Rice. With Bichowsky Urey continued working on the Zeeman effect (the splitting of a spectral line in the presence of a magnetic field), a topic brought to his attention by Kramers in Copenhagen, and published one paper on the possible magnetic qualities of a spinning electron.[32] However, Urey came to regret publishing the results of this latter collaboration.

Urey and Bichowsky's paper followed on the heels of an earlier contribution by two graduate students from Leiden University, George E. Uhlenbeck and Samuel A. Goudsmit, in which they first proposed the concept of electron spin.[33] Urey and Bichowsky asserted in their own contribution that the idea had occurred to them "quite independently and for largely the same reasons," but that they had "carried the idea somewhat further" than Uhlenbeck and Goudsmit.[34] Urey later came to question whether he and Bichowsky really had gone beyond their peers, and felt sorry that their competing claim on the concept might have prevented Uhlenbeck and Goudsmit from winning the Nobel Prize. In an interview, Urey admitted that his decision to publish had been based on his own ambitions:

> I've always been a little bit sorry we published it, because I think it prevented a Nobel Prize for the spinning electron to Goudsmit and Uhlenbeck. I always feel sorry about it, and I wrote to the Nobel Committee telling them so. Because I don't think we added much beyond what Goudsmit and Uhlenbeck did, maybe we only added confusion. It was a completely original idea with us, this I always insist on, but I'm sorry that we just didn't shut up. It was a matter of young people, you know, as I often say, trying to get ahead, and so forth. If I had been a little older and a little more mature I wouldn't have done it.[35]

With Rice, Urey began producing his first graduate students. Together they co-advised two doctoral dissertations on the mechanism of homogeneous gas reactions, one concerning blackbody radiation and its effects on a molecular beam of nitrogen pentoxide and the other on the absorption spectrum of this same gas.[36] On his own, Urey advised two doctoral dissertations that dealt with the properties of atomic and molecular hydrogen.[37]

Urey's fear that Hopkins would continue to be an inhospitable place for interesting research seemed to be confirmed when Ames assumed the presidency in 1929. By Urey's account, this began a thirty-year decline in the university's excellence, as he and several of his colleagues left to find universities where their talents might get them promoted. Urey had no qualms about telling Heilbron that "this man destroyed the university." When Columbia University offered Urey the opportunity to leave Hopkins to take up the post of associate professor of chemistry, he leapt at the chance. He wrote to Birge that he was eager to give up Hopkins for "the interesting group of men in New York City."[38]

COLUMBIA AND THE DISCOVERY OF HEAVY HYDROGEN

Despite the improvements Urey made in the teaching of theoretical and mathematical chemistry, Hopkins was not Berkeley. Urey found himself missing the robust experimental facilities of Gilman Hall, and later admitted that he felt "foolish" not to have gone back to California where he could have picked up his dissertation research once again and brought it in line with the new quantum mechanics.[39] This would have required applying for a National Research Council fellowship at Berkeley rather than Harvard, which Urey seemed unwilling to consider at the time. While much of Urey's frustration at Hopkins had to do with being in a department that was not interested in thermodynamics and heat capacities, it also had to do with the university's lack of facilities within which he could perform the type of low-temperature experiments that this work would require. This was at least part of the reason that he continued working on theoretical problems for so long after he had already recognized his limitations.[40]

Once Urey had moved to New York, the resources available to him as an experimental physical chemist increased greatly. Urey wrote to Lewis that his new research facilities were "the best that I have seen since I left California. What a lot of help it is to have good physical facilities."[41] He told Birge that "apparatus, which requires weeks to get in Baltimore, and much correspondence, can be secured here by telephone; also the place has much more money than Hopkins, which, as you know, is a distinct advantage."[42] Of particular interest to Urey was an underused spectrograph:

> In a way, the transfer to Columbia University seemed to me a more likely place for extensive development than Johns Hopkins. It was fortunate in a way, because at Columbia there was a spectra apparatus which was not being used by anybody, and which I could use for studying the spectrum of

> hydrogen and discovering heavy hydrogen. If I had remained at Hopkins, no such apparatus would have been available to me, since Professor Wood [in the physics department] was using this apparatus continuously.[43]

At Columbia, Urey also now had a dedicated research assistant, George Murphy. In the basement of Columbia's Pupin Hall, Urey and Murphy modified the "spectra apparatus" Urey mentioned into a twenty-one-foot grating spectrograph of their own design.[44] While they would eventually use this apparatus to discover heavy hydrogen, they first refined it by studying the relative abundances of recently discovered isotopes, such as those of nitrogen and oxygen. Their driving question at this time was whether these abundance ratios differed in samples of different chemical origin.[45]

Urey's scientific output increased significantly at Columbia. From one or two papers per year while at Hopkins, Urey published seven papers in 1931 alone. In part this increase was related to the number of collaborative projects and dissertations in which Urey was involved. In addition to his work with Murphy, Urey had many other irons in the fire. He was studying the absorption spectra of chlorine dioxide with a doctoral student, Helen Johnston, and finding that isotope effects were helpful in the analysis of the observed spectral bands.[46] With Charles Bradley he was studying the recently discovered Raman effect, and using this phenomenon to determine the normal vibrational frequencies of the polyatomic molecule silicochloroform.[47] As many of these publications were released in or around 1931—the same year that Urey, Murphy, and Brickwedde announced their discovery of deuterium—we might consider this Urey's annus mirabilis. The work done during this period, along with his earlier work on gases at Berkeley, certainly did position Urey as one of the world's leading experts in statistical and spectroscopic chemistry.

His work was moving him in the direction of isotopes, and it was in this subfield of physical chemistry that Urey would achieve his lasting fame. He was nudged in this direction by one of his former colleagues at Berkeley, William Giauque. Giauque had earned his doctorate under Gibson in 1922, and Lewis had quickly offered him a faculty position within his department. Like Urey, Giauque had also developed a relationship with Birge and abandoned any strict separation of chemistry and physics. When Urey left Berkeley in 1923, Giauque took over his work on the entropy of gases and, using Berkeley's low-temperature facilities, did the type of work on heat capacities that Urey complained he could not do at Hopkins.[48] In 1929, just as Urey made his move to Columbia, Giauque and Johnston published a series of papers based on their work on the heat capacity of liquid oxygen from 12°K (−261°C) to its boiling point.[49] Two of these papers reported isotopes of oxygen of masses 17 and 18. These discoveries held implications for possible isotopes of hydrogen.

Work on isotopes was relatively new. In 1913, the same year that Bohr introduced the world to his atomic model, the radiochemist Frederick Soddy proposed and the physicist J. J. Thomson experimentally confirmed the existence of different types or species of atom occupying the same place on the periodic table and differing only in mass. By 1919 the British physicist Francis W. Aston had constructed a mass spectrograph at his Cavendish laboratory that used magnetic and electric fields in order to separate and measure isotopes by their atomic weights. Aston had previously measured the atomic weight of hydrogen with his instrument and found a value that strongly agreed with the value determined by chemical means.[50] Aston's method, however, had assumed a standard atomic weight for oxygen of 16. With Giauque and Johnston's announcement of the two new isotopes of oxygen, "it was necessary that there be a heavy isotope of hydrogen."[51]

As isotopes were understood at the time, not many physicists or chemists believed that a heavy isotope of hydrogen could be found; they considered it unlikely to exist. The concept of the neutron had yet to be introduced, and so the nucleus of the atom was at this time thought to be composed of protons and nuclear electrons. Isotopes were understood to differ in mass because they possessed differing numbers of these two nuclear components. An additional electron in the nucleus of hydrogen—understood in normal hydrogen to consist of one single proton—would have negated the hydrogen nucleus's charge. Even after Giauque and Johnston's work led to the reconsideration of Aston's measurements, the heavy isotope's discovery seemed like an incredible challenge. Calculations by Birge and the astronomer Donald Menzel at Berkeley indicated that if the heavier isotope did exist, it would be exceedingly rare—composing only about one part in 4,500 in naturally occurring hydrogen. But Urey had already suspected the existence of deuterium, based on a chart he had constructed in his office depicting the possible arrangements of protons and electrons in nuclear structure. The chart had as its abscissa the number of electrons and as its ordinate the number of protons. Looking at his chart, Urey noticed that the known nuclei of light elements conformed to straight line segments within which would fit a heavy isotope of hydrogen.

Urey and Murphy were confident that if the isotope existed, and if they could enrich a sample of hydrogen with its heavy isotope, then they could use their spectrograph to detect it. Urey devised a method of enrichment that exploited the thermodynamic qualities of hydrogen and the theoretical differences in vapor pressure of its isotopes at their triple point.[52] The experimental design drew on his earlier graduate work on the heat capacity and entropy of gases, which led him to predict that there would be a difference in the vapor pressure of the isotopes in liquid hydrogen.[53] He hypothesized that distilling five-liter quantities of liquid hydrogen down to a residue of only two cubic centimeters of liquid at the correct temperature and pressure would produce a several-fold increase in the concentration of the heavy isotope. In the early 1930s, however, there were only two laboratories in the United States that could reliably achieve the 20.28°K (−252.87°C) temperature required to produce liquid hydrogen in large quantities, let alone distill the liquid at its triple point of 13.84°K (−259.31°C). One such lab was at Berkeley, where liquid hydrogen and air plants had been installed in the Gilman Hall basement and subbasement. These were the facilities used by Giauque.

The other suitable laboratory was at the National Bureau of Standards. Urey called on Brickwedde, his old seminar colleague and chief of the bureau's low-temperature laboratory. Brickwedde had a long-standing interest in atomic structure and had even written his master's thesis at Hopkins on Bohr's atomic model.[54] When Urey approached him and outlined his experimental design, Brickwedde agreed enthusiastically to the collaboration. Brickwedde's equipment was in need of repair and reassembly, and so it would be several months before the enriched hydrogen samples could be produced and analyzed.

In the meantime, Urey and Murphy went ahead with their experiment using a commercially prepared tank of hydrogen gas. They were surprised to find that they could detect the predicted spectral line of the heavy isotope even from this unenriched sample. While the spectral line of normal hydrogen appeared on the plate after an exposure time of only one second, a fainter line in the predicted location of H-2 appeared after an exposure time of one hour.[55] Urey was excited by this positive result, but he decided not to report his findings immediately. He and Murphy bided their time, eliminating possible sources of error in their apparatus and methods. When they finally did receive the enriched samples from Brickwedde, they found the H-2 line with an exposure time of only ten minutes. They therefore concluded that the H-2 lines they had seen in their earlier runs truly were H-2 lines.

The discovery of heavy hydrogen—which Urey named deuterium—stimulated a flurry of research around the country. Urey's colleagues estimated that between 1931 and 1934 more than two hundred papers concerning deuterium appeared in print.[56] Lewis, who managed to be the first to isolate a highly concentrated sample of heavy water, was one of the most significant contributors to this literature, writing more than twenty-five papers on deuterium and heavy water during these years. According to Lewis's student and research assistant, Jacob Bigeleisen, "Lewis jumped on the bandwagon."[57] He was able to work so rapidly on the problem of heavy water because he already had a large store of enriched water from the electrolytic cells Giauque had used when generating hydrogen gas for liquefaction in his earlier experiments.

Urey began to resent Lewis's sudden activity in deuterium studies, and felt that his former teacher was unfairly attempting to outpace him using the incredible advantages of his superior chemical facilities and reputation:

> [Edward W.] Washburn was, I thought, somewhat slow in developing the electrolytic separation of the hydrogen isotopes, and it seems that Professor Lewis . . . also felt that Washburn and I were rather slow about this, and so he undertook to prepare heavy water pure, and of course was the first one to prepare pure heavy water. Washburn and I were working on it with considerably less facilities at our disposal than the chairman of an enormously important chemical department in the United States.[58]

Moreover, Urey felt slighted that, even with his great discovery in hand, Lewis had not invited him to join him at Berkeley but instead seemed to be trying to steal his thunder. As Urey wrote to Birge in the 1960s, "It would have seemed to me that in a similar situation one might have thought [Lewis] would have invited me to come to California and would have helped me to develop the work and taken pride in a former student instead of somehow trying to take credit for himself."[59]

Urey's letters to his Berkeley mentors from this time were not silent about his discomfort with Lewis's jumping into heavy hydrogen work. When Lewis submitted a paper on deuterium enrichment to Urey's *Journal of Chemical Physics*, Urey wrote back that this put him in an awkward position. He knew that Washburn had developed the electrolytic separation method prior to Lewis, and had conscientiously waited for Urey to complete the separation of hydrogen isotopes. Seeing that Lewis had entered the race, Urey felt compelled to accelerate his own separation work and to push Washburn to work faster.[60] When Urey heard that the Berkeley chemists were spreading the rumor that Urey was patenting a separation method that Lewis had developed, he wrote to Birge to deny emphatically that there was any truth to this: "I have never considered the possibility of patenting the process. . . . The only suggestion in regard to patenting the process which I made was to Dr. E. W. Washburn who discovered the method—not G. N. Lewis. I think it would be quite foolish for me to undertake to take out patents on a process discovered by another man because those patents would not be worth the paper they were written on."[61] Lewis's interest in heavy hydrogen and heavy water came to an abrupt end in 1934, when Urey alone won the Nobel Prize for his discovery of deuterium. Lewis, it seems, had worked under the impression that he and Urey might share the prize (in fact, rumors that the prize would be shared were prevalent in the days leading up to the announcement). In addition to ceasing all work on deuterium, Lewis published nothing at all for the next eighteen months and resigned from the National Academy of Sciences. Lewis soon estranged himself from Urey, who struggled for years after winning the prize to re-ingratiate himself to his former professor.[62]

That Urey expected to be welcomed back into the Berkeley fold after his discovery of heavy hydrogen is also suggested by his correspondence with Kenneth Pitzer, one of Lewis's recruits and his successor as chair of the College of Chemistry. In the 1960s, Urey wrote to Pitzer complaining that his own contributions to the thermodynamic properties of deuterium and methods for its separation were never included in the revisions to Lewis's textbook on thermodynamics. Urey noticed that both he and Irving Langmuir (whose contributions to refining the Lewis-Langmuir atomic model Lewis had never appreciated) were largely absent from the text and wondered why Lewis treated them as "outcasts."[63] In another letter to Pitzer, he wrote, "It is curious to me that I have never been accepted as part of the honored graduates of Berkeley. It has gone on for close to 40 years. Probably my fault. But why?"[64]

An excerpt from Urey's interview with Heilbron sheds some light on the possible reasons for wanting to leave Columbia and return to Berkeley, aside from his desire to be welcomed back into the fold:

> Columbia has been kind of a dead dull place, in chemistry particularly; and it has been that way ever since the turn of the century. Somehow a university gets a certain tradition, and you just can't change that. . . . [There was] a great deal of personal jealousy between people in the department instead of friendly boosting of each other. At Columbia I had very few friends. They weren't really friendly to each other. It wasn't just a matter of the outsider coming in from Hopkins, being the ugly duckling that everyone picked on. It wasn't that. They weren't friendly to each other, those that were there. Or the people who have been there since.[65]

This lack of collegiality in the department would become even more apparent to Urey as department chair between 1939 and 1942, a time he later looked back on with no fondness.

In addition to the faculty's unfriendliness, Urey also felt politically at odds with the rest of his department, which was noticeably Republican. Urey's humble roots gave him a different outlook on the world than his elite colleagues. This became clear during the 1936 presidential election between Franklin D. Roosevelt and Alfred Landon. Urey believed in the New Deal and took on the role of political activist during this period, sponsoring such organizations as the American Association of Scientific Workers and speaking publicly on the importance of sharing America's wealth and its burdens. In Peter Kuznick's account of this period, he identified Urey as "clearly [representing] the left wing of the American scientific community" and as a leader in the movement for social responsibility.[66] But in Columbia's Department of Chemistry, Urey's colleagues did not share his political views. As one of Urey's graduate students from this period, Mildred Cohn, remembered, "In 1936 when Roosevelt was running against Landon, Urey sported a Roosevelt button. But he was the only member of the chemistry department who did. The others all had Landon buttons."[67]

Urey also felt that the department was intolerant of anyone who didn't fit the norm of white Protestant male. This was made especially apparent to him by the experiences of his Jewish graduate students. Urey discovered more than deuterium during the 1930s; as Cohn put it, "Urey discovered anti-Semitism."[68] While Cohn insisted that Urey had never known anti-Semitism before moving to New York, he became a quick study. Having grown up and attended college as something of an outsider to Anglo-American culture, he no doubt saw something of his own struggle in his Jewish students' experiences. Many of his graduate students during the 1930s were Jews who had grown up in New York City or on the East Coast and, despite anti-Semitism and Jewish quotas on enrollment, attended Columbia because they did not want to leave their families.

Rabi, who in 1929 was the first Jewish physicist hired at Columbia, recalled that it was very difficult for Jews to get university jobs in the 1920s and '30s. While many advisers were willing to take on Jewish students and assistants in their laboratories, they did not do much to help place them.[69] The situation was no different at Columbia, where Rabi felt he was only hired because Heisenberg had recommended him for the job (a job for which Rabi had not applied because he felt there was no hope) while on an American lecture tour.[70] Cohn remembered similarly. Discussing the case of one potential Jewish faculty hire, Cohn said, "It never came to pass, and it was obvious to everyone concerned that the reason was because he was a Jew."[71] Cohn also remembered that a similar situation faced women in the sciences: "I have talked to other women of my generation, and they tell me that even though their professors took them on as graduate students, they never really expected them to have careers, particularly if they were getting married."[72] But Urey was different: "Urey never took that view. He assumed that I would have a career whether I was married or not. . . . [From] that point of view [I was lucky], because I know women who told me that their professors didn't bother trying to get them jobs."[73]

Urey soon gained a reputation among Jewish graduate students as someone who would fight for them within the department, help them to find support during their graduate research, and help them find jobs once they had graduated. Cohn, one of the few Jewish women to attend Columbia in the late 1930s, remembered Urey as "the only professor in the chemistry department at Columbia who was concerned with the welfare of the graduate students in those Depression years. Were they paid enough as teaching assistants? Were the long hours they worked interfering with their research?" In order to protect Cohn from long hours of work outside of the lab, Urey offered her a loan, telling her, "Ever since I got the Nobel Prize, I've wanted to use some of that prize money to help my students. So, why don't you let me lend you some money, and some day when you have a job, you can pay me back."[74]

Still, Urey didn't believe that his support alone would make up for the department's antagonism toward Jewish students. He advised Cohn at one point that she might escape prejudice by moving somewhere like the Midwest, where he believed anti-Semitism was not such a prominent part of academic or social life, and where she might reinvent herself through marriage to someone not Jewish. She recalled:

> And at one point he said to me, "You know," he said, "why don't you go out to the Midwest where there's no prejudice and marry a non-Jew and forget that you're a Jew? Then you won't have these problems." So I said, "Has it ever occurred to you that maybe I don't want to forget that I'm a Jew?" He was genuinely surprised that one wouldn't want to get rid of such a handicap.[75]

His own experience had taught him that one could "forget" those things about oneself that attracted the prejudice of others. But he soon came to believe that Jews could not transform themselves in quite the way that he had. Rather, his identity was malleable in a way that some others were not. As he told Cohn of Rabi, "I can conform but he cannot."[76] Behind closed doors he even advised David Altman, a Jewish student from Cornell who had been accepted to both Columbia and Berkeley, that he would do better to avoid the intolerance of Columbia.[77]

Urey's generosity was not limited to his students. Although he alone won the Nobel Prize for the discovery of deuterium, he chose to split the prize money equally with his two collaborators. He also shared money with unsupported colleagues whose work he admired. Shortly after winning the Nobel Prize, Urey was awarded a $7,600 research grant from the Carnegie Institution with no strings attached. Half of this he gave to Rabi, who would go on to win the Nobel Prize in 1944. As Rabi remembered, "I had had nothing to do with his discovery [of deuterium]. What a greatness in Harold Urey—what a tremendous magnanimity to do something like that! He had a deep faith in me. When he came back from receiving his Nobel Prize, he told somebody, referring to me, 'That man is going to win the Nobel Prize.' . . . This money set me free. It made me independent of the Physics Department."[78] It is possible that, in addition to believing in Rabi's work, Urey also saw something of himself in Rabi. Like Urey, Rabi came from a very pious family and occupied two worlds because of it. While in the outside world Rabi was a secular scientist who put little stock in God as anything more than a useful "heuristic principle" in understanding the mysteries of the physical universe, within his Orthodox family at home, Rabi "was a good son," conforming to his family's views and showing a genuine respect for the traditions of his ancestors.[79] As political tensions around the world began to grow in the coming decades, Urey may have longed for the type of private devotion Rabi enjoyed.

CHAPTER FOUR

From Nobel Laureate to Manhattan Project Burnout

After winning the Nobel Prize, the 41-year-old chemist had the ability and opportunity to pursue whatever lines of research he chose. In his family life, he was more than satisfied. He and Frieda now had three daughters. He had skipped the Nobel Prize ceremony because of the birth of the third, Mary Alice. Frieda seems not to have played much of a role in the daily activities in his lab, but she was Harold's constant companion when he traveled to conferences or speaking engagements. In February 1935 the *New York Times* reported that she would accompany him to Sweden, where he would finally appear before the Swedish Royal Academy of Science and deliver his Nobel address.[1] In 1939 they had their fourth and final child, a son they named John Clayton.

Frieda by now had discovered a strategy for those times when Harold seemed to grow distant from her or the children, lost in his own thoughts: she would give him a kick to bring him back to reality. The children learned that they could get his attention by asking him about dinosaurs, at which he would launch into an explanation of paleontology. They could also draw him into sing-alongs of Gilbert and Sullivan numbers. On more serious occasions, he taught them the ethical beliefs he had learned as a child.[2]

With increased resources from external grants, Urey was now able to support a small team of graduate students and postdoctoral researchers, and the pace of the small lab's work accelerated. Much of his work for the remainder of the 1930s followed the plan he laid out in his Nobel address.

FIGURE 9 Harold and Frieda Daum Urey at the May 1939 meeting of the National Academy of Sciences, Washington, DC. Lyle H. B. Peer is in the center. Smithsonian Institution Archives, Science Service Records, image SIA2010-0260.

With his graduate students he explored the chemical differences between hydrogen and deuterium and their respective compounds. Using spectroscopic data, his lab was able to calculate the changes in nuclear mass, spin, entropy, and free energy that resulted from the substitution of hydrogen and deuterium. Urey also worked with his graduate students and research assistants to calculate the exchange reactions involving the isotopes of the other light elements. Once they had calculated the equilibrium constants for the exchange reactions, they were then able to develop new chemical methods of isotope enrichment using distillation columns. Urey and his team—which from 1934 to 1939 included John Huffman, Clyde Hutchison, David Stewart, and research assistant Harry Thode—began distilling and fractionating high concentrations of the rare isotopes of carbon, nitrogen, oxygen, and sulfur.[3]

With a separate group of colleagues and students—Irving Roberts, Mildred Cohn, and Isidor Kirshenbaum—Urey began putting these isotopes to work in research projects, including the study of chemical reaction mechanisms and the differences in vapor pressure of isotopic compounds. But in addition to the work being done in his own lab, Urey was also providing enriched isotopes to biochemists at Columbia and at universities around the country for work on metabolism and body chemistry. The biochemist Rudolf Schoenheimer arrived at Columbia in 1934 and set about finding ways of using deuterium-enriched compounds (as well as other stable isotopes) acquired from Urey to investigate the sequences of metabolic reactions.[4] By any standard, Urey's research program was highly productive.

As a Nobel Prize winner, Urey was a highly sought-after public speaker. His earliest speeches, delivered in the context of the Great Depression, were delivered partly in defense against Depression-era attacks on science. American science had entered the 1920s on a high note, having contributed to victory in World War I and in the process developing strong ties to American industry. Chemistry's claim to cultural relevance was thus bound in no small part to its ability to raise American industrial strength to a level that at least matched that of Europe.[5] Within their own publications in the early 1920s, the American chemists congratulated one another for having demonstrated their self-sufficiency and centrality to the American economy.[6] Now, in the early 1930s, critics wondered if science and technology had not grown American industry and changed the labor market too quickly.[7] As Kevles points out, some of the harshest attacks on science were often articulated by religious critics who saw faith in the improving power of scientific research as faith that would be better placed in religion.[8]

During the 1930s, Urey did not agree. From World War I and his training at Berkeley, he had inherited a professional ethic that combined a research ideal imported from Germany but tempered in American industry, with a public service ideal developed during America's Progressive era. According to the historian Gilbert Whittemore, the American chemists' research ideal "encompassed both pure and applied research, with the line between the two becoming less and less well-defined."[9] The profession's public service ideal held that research was not only a means to an end in the laboratory or factory, but also "a powerful tool for the general welfare," and that science was "not the physical laws governing nature, but a procedure or an orientation which could be applied to the problems of society."[10] The strongest interpretation held that the chemist was the quintessential "public man," equally qualified to conquer technical or social problems with his rational methods and his ability to reshape and reorganize the physical world.[11] Urey embraced the role of public man.

Urey was certainly not blind to the problems of the Depression, nor did he take the scientific community's responsibility in solving these problems lightly. His speeches made it clear that he believed science to be much more than a utilitarian pursuit; along the lines of the public service ideal, he believed that scientists had a clear social responsibility to improve the human condition, even when scientific discovery was not the direct cause of human suffering: "After all each of us has but one life to live and that life is our most valuable possession. We should not overlook and excuse the sentence of poverty, privation and disappointment on an innocent fraction of the population on the grounds that the result is good for the greatest number, nor should we excuse our responsibility in the matter because of the indirectness of the method."[12] While he reminded his audiences that "it is probably true that people live better today in both an intellectual and spiritual as well as a physical sense than they ever have before in all history," he also admitted that civilization still had many imperfections, "emphasized by many discourses and more by much discomfort and disappointment."[13]

Urey was even willing to admit that the rapid advance of science and technology might have something to do with the economic and social woes of the Depression. While in the past industry had advanced at a pace that allowed workers time to adjust to any changes in the demand for labor or the skills required to work in local industry, an unfortunate "by-product" of the application of science to industry during the second industrial revolution was that "industries are often born, grow to maturity in a year or so and are then often murdered. The change is so rapid that people cannot adjust themselves to the change and even though the end result is beneficial, the intermediate situation may result in actual want and privation for many people."[14] No doubt drawing from his own personal experience on his family's unsuccessful farms in Indiana and Montana, Urey described the problems that the introduction of synthetic fabrics might pose to cotton farmers in the south, noting that it could take an entire generation to adjust a region's agricultural practices to a new crop.

To the ideals of his profession, Urey's experiences in Europe had added a critical view of American capitalism. Although he never directly invoked socialism in his Depression-era speeches, he did come close. In Copenhagen, Urey had learned not to fear the welfare state: "There was a country with many socialistic things that have been practiced for many years, and the country was as free and democratic as my own."[15] Urey speculated that "some fatal defect" in the American economy had caused its breakdown in 1929, and that one of the greatest contributions chemistry might make to society would be "to profoundly modify our social and economic institutions" to match the "greatly increased productive capacity" that the practical application of chemistry made possible.[16] World War I and the Depression, he believed, were both primarily the results of old economic relationships that had envisioned neither mass production nor the modern interdependence of the nations of the world. Urey understood history to be a story in which "the general progress has been toward the left," and expected "that by the year 2000 very marked changes must come in our social structure."[17] He encouraged one audience of graduating college seniors "to take with you some idea and philosophy of life that will enable you to transcend selfish advantage which you might secure, and moreover, I should like to suggest that you take an open mind toward social changes which in any case are inevitable, and which we should all welcome rather than resist."[18]

Even as Urey was willing to consider the social and economic problems of the Depression as indirect consequences of science applied to industry, in the mid-1930s he was also quick to point out that since the end of the war, "no country [had] arranged its internal affairs in such a way that anything approaching the maximum production of material goods for peaceful purposes [had] been accomplished."[19] Just as the desire to win the war had been matched by a productive and efficient war industry, the desire for peace and "for the good things which science and technology can bring" must be matched by an active "production machinery": "Our people wish those things which chemistry can bring to them, but for some reason our chemical plants are partly idle, our chemists unemployed, and our workmen are on some form of direct or indirect relief."[20] This was a problem that chemistry could not solve. He hypothesized that the true cause of the problem, as well as of "most of the serious difficulties of this century"—including the Great War, communism in Russia, and the worldwide depression—were caused by "the great production of goods by scientific methods, together with archaic methods of distribution."[21]

Still, Urey was hopeful that the increased productive capacity made possible by science would modify the social and economic institutions of Western countries in beneficial ways. He told his audiences that, if properly applied, science and technology could aid in the elimination of war and poverty and extend comfort and wealth to all social classes of all nations. If mass production was to continue, Urey argued, it must be matched by mass distribution and consumption: "Whatever that distributive system will be, it must distribute an abundance to many people and not to the privileged few only. If this is not done, we must abandon these mass production methods."[22] With the application of scientific reason to the problem, Urey was confident that society could emerge from the Depression and enter a new and more economically just stage in its development. As was often the case in Urey's speeches, his audience was faced with two mutually exclusive visions of the future: "If we act with courage our descendants will live in an abundance of necessities and luxuries the like of which we can not imagine. If we are not courageous, they may live with less than we have at present."[23]

As for the future of science, by the late 1930s Urey was convinced that two fates were possible. One was the further militarization of science. Here—foreshadowing his postwar concerns over atomic weapons—he imagined that science's support of military activity could bring about "the complete destruction of our civilization."[24] Although scientists had lofty ideals, they could be diverted from these by "the stimulus of the belligerent and destructive human instincts."[25] This had been the case for German science leading up to World War I—and it had led to the development of work on the fixation of nitrogen for military purposes. He warned his colleagues at the American Association for the Advancement of Science that chemistry, contributing as it had to the design and construction of ever more explosive and effective weapons, "can and perhaps will destroy our European civilization."[26] He speculated that another world war could not only destroy the wealth of nations, but could also "damage governmental regimes beyond repair and can make a world less safe for democracy or anything else."[27] But there was also the possibility that, with the right encouragement, scientists could instead work toward peaceful ends. Along these lines, Urey suggested to his colleagues that they focus their activities on discovering chemical substances "which would stimulate such creative endeavor for the arts of peace."[28]

Because science could and often was diverted by the interests of the society within which it was practiced, Urey did feel that science prospered best when practiced in the proper political and social context. Science seemed to have blossomed in countries that had embraced liberal government and the development of humanism. Indeed, he suggested to a Pittsburgh audience that "the center of gravity of the sciences in the future will move with the center of gravity of liberal government."[29] This was perhaps why chemical research—so firmly established in German industry prior to the war—had now moved toward North America. Urey did not articulate any reasons for why science should align itself with liberal government, or for why his audiences should expect to see Western democracies embrace further social change. These things he took for granted as natural conclusions.

Urey was not alone in seeing a link between science and liberal government.[30] The historian David A. Hollinger has noted that during this time a large camp of antifascist thinkers pointed to science as a product of proper political culture, specifically democracy.[31] The sociologist of science Robert K. Merton, for example, who made much of the relationship between science and social structure, rooted his 1942 work on science and democracy in just this idea. Merton's position was not unique, influenced as it was by Max Weber's thesis that claimed a link between the Protestant ethic and the emergence of capitalism. According to Hollinger, Merton's ideas stand as "a benchmark in the emergence of social definitions of the scientific enterprise and in the development of ideological self-consciousness on the part of apologists for science."[32] Urey may be considered one of these apologists.

SCIENCE AND RELIGION IN THE INTERWAR YEARS

In these early speeches, Urey did not show much sympathy for his religious critics' insistence that too much faith had been placed in science. There is no indication that he felt any need to defend the traditional moral teachings of religion. Instead, he framed his description of the chemical and scientific profession in explicitly religious language, and at times claimed outright that science was a religion. In a speech delivered at the dedication of the new buildings of the Mellon Institute of Industrial Research, he told his audience that the "real purpose" of scientific activity was "to contribute something somewhere and at sometime to the sum of human satisfaction, as man lives for a brief span of time on this small planet."[33] While much of what the public would recognize as chemistry's contributions to life were felt in the physical realm, the scientist's highest aim was to contribute to the spiritual and intellectual satisfaction of humankind. In its applied form, chemistry contributed to these higher pursuits by reshaping the physical world of humans in such a way that it allowed them to transcend the animal world, freeing them from preoccupation with their physical needs and wants. The ultimate aim of applied chemistry was to "abolish drudgery, discomfort and want from the lives of men and bring them pleasure, comfort, leisure and beauty."[34] In its purer form, chemistry and its sister sciences helped broaden humankind's intellectual horizons.

Like the Dunker faith, this religion of science was defined not in words but in actions, and was thus meaningful only to its practitioners. As for those who practiced the religion of science, Urey assured his audiences that he and his peers were special members of the community. Their primary objectives were not "jobs and dividends," and they were "satisfied with modest salaries," knowing that their work would improve the human condition. In a moment of bravado, he told the crowd in Pittsburgh, "You may bury our bodies where you will, our epitaphs are written in our scientific journals, our monuments are the industries which we build, which without our magic touch would never be." He went on to suggest that the plants where chemical processes were applied to industry should indeed "be regarded as national monuments," although he admitted that "they shun the public eye, are located in small and isolated towns often with dingy surroundings and no multitudes make yearly pilgrimages to these Meccas."[35]

But the satisfaction of improving the lives of others through their contribution to industry was not the greatest satisfaction for the scientist; nor was it the only reason the scientist was willing to sacrifice personal wealth. In one of Urey's earliest public addresses after winning the Nobel Prize, delivering the commencement address at his alma mater, the University of Montana, he introduced a higher form of satisfaction: direct communion with the laws of nature. He compared the work of the scientist to that of other intellectual workers who "find [themselves] engaged in a search for truth—let us not try to define it too precisely—as well as the beautiful."[36] This search for truth and beauty demanded more than a sacrifice of personal wealth and well-being; it required complete supplication to the methods of science and objectivity. These searchers were defined by their willingness to regard their subject matter above themselves, and to make whatever sacrifices were necessary in order to take a disinterested and unprejudiced position. Thus Urey defined objectivity as a type of transcendence born of sacrifice—as an "attempt to dissociate ourselves from the subject and to consider it from a somewhat higher plane than our own individual desires, ambitions and fortunes would dictate."[37] In his 1938 address to the American Association for the Advancement of Science, the rewards of this sacrifice and transcendence took the form of communion: "If [the scientist's] postulates are in accordance with what we call natural law, nature unlocks her secrets with an amazing ease. When this occurs there is a communion between scientists and the eternal laws governing the behavior of this universe that is very intimate indeed. This communion represents the highest reward which a scientist receives for his services, and it is this that furnishes the major driving force for all his activities."[38]

While Urey did in these early speeches address scientists' ethics, he did not take the position that would define his later Cold War speeches—that scientists were ethical only because they had been raised in a religious society. Rather, he took the position that one useful application of science to everyday life would be to educate the public about science's ideals and ethics. He noted that, regarding the sciences and the humanities, "these fields are dominated by an honesty of purpose, a sincere desire to find the best." Drawing on his discussion of the sacrifices and the transcendence of research, he suggested that the scientist and the humanist were able to "submerge" their own individuality in a more important purpose. While they may remain concerned to an extent about their own personal well-being, their demands were rarely beyond reason. And their ethics with regard to the truth were beyond reproach:

> None of us would hold to an incorrect theory or experiment of our own in the face of proof to the contrary. We would not perpetrate a lie or an inaccuracy on our fellow-workers or on those to follow us, for our own selfish ends. Largely throughout all our science we have eliminated emotionalism. We do not try to win our debates by sarcasm or by any other tricks which might gain our immediate ends, but which would give us an inaccurate value of nature, and what is more, we will not tolerate any such atti-

> tude on the part of other scientists. If scientists do those things, they soon become complete pariahs whom all ignore. It is this sincerity of purpose and attitude found among the scientific groups, and also existing in the humanities, which I believe is the most important lesson science brings to these other fields of work.[39]

Thus the scientist was a shining example of moral integrity. This was a result of sublimation to the subject, not of anything brought into the scientific profession from the surrounding culture.

Aside from science being a type of religious experience for its dedicated practitioners, there was little space for religion in Urey's early rhetoric. Channeling the freethinker Robert Ingersoll, perhaps, he asserted that the progress of science had "driven clouds of superstition away" from human minds:

> In the past many of our sister sciences have contributed notably to the broadening of man's intellectual horizon. Astronomy gave him a proper perspective of his own importance and his own position in this universe. His earth is not the center of the solar system, nor is his solar system the center of this galaxy. At the same time that he was robbed of his central position in the physical universe what grandeur has been spread before him as our knowledge of astronomy grows! Biology has robbed him of his miraculous creation by one or another anthropomorphic god, and has placed him in the lowly position of one of the animals, that one, in fact, who in this particular geological age dominates the earth, but at the same time the grandeur of an organic evolution with time has been spread before him. Geology has shown him the long length of time that man and his lowly ancestors have existed on this planet, and in fact that life on this planet has been nearly coexistent with the planet.[40]

Chemistry, meanwhile, had made the world more volatile but also more wonderful. There was no indication in Urey's early speeches that he felt these changes from traditional ways of life would have negative consequences. He seems, at least in the 1930s, not to have cared much at all if science robbed man of his gods. If anything, this removal of "superstition" was part of the progress Urey was celebrating, and it was a small sacrifice in the attainment of the grandeur of the narrative that was being produced. It was rational choice, not traditional morals, that would transform the world into one of leisure, abundance, and peace.

Urey's position on the relationship between science and religion began to shift in the late 1930s, as war in Europe again loomed on the horizon. Evidence of this is found in his participation in an intellectual project intended to apply the method of "corporate thinking" to "see what scientists, philosophers, and theologians could do to unite the more abstract thought and thinkers of the present in defending democracy."[41] This was the Conference on Science, Philosophy, and Religion in Their Relation to the Democratic Way of Life, hosted at Columbia University and sponsored by the Jewish Theological Seminary.[42] The conference was only one of the many manifestations of the preoccupation with fascism and totalitarianism among New York intellectuals during this time.[43] However, it also represented a special concern over the increased specialization of intellectual work and the "departmentalization of American thought."[44] One of the central tenets of the conference, in the words of one participant, the historian Van Wyck Brooks, was the recognition

> that our failure to integrate science, philosophy and religion, in relation to traditional ethical values and the democratic way of life, has been catastrophic for civilization. . . . We know that democracy exalts the individual but that individualism as an end in itself means anarchy. We know that tradition can make slaves of men, but that the lack of historical perspective and rootage in the past makes their lives thin and unheroic. We know that technology can be and has been a great emancipator, but that it may also put tools in men's hands with which to destroy their heritage.[45]

The aim of the conference, Brooks argued, was "to promote respect and understanding between the three disciplines involved and to create among them a consensus concerning the universal character of truth."[46]

The group was especially concerned over the threat of totalitarianism. The conference adopted early on a model of totalitarianism within which the ruler derives authority because "the primitive identification of the state and the Deity has been re-established."[47] This "pseudo-religious" philosophy was incorporated into every part of the totalitarian empire's government, economy, and society, making possible the "worship of power, and the contempt of truth, mercy, and justice," and leading the followers of Hitler, Stalin, and Hirohito to worship them as "quasi-divine personages."[48] The group feared that the totalitarians might take advantage of "decreasing respect for ethical and religious values among the democratic peoples" and the resulting "confusion in their educational systems, in their literatures, and in organs of public opinion generally."[49] They worried that "a cynical, divided, hyper-individualistic America will necessarily become a doomed America," and offered as remedy cooperation between leading scientific, philosophic, and religious thinkers. This, they hoped, would produce a "dynamic philosophy of American democratic life" that would "oppose any effort at deification of the state, or the suppression of individual liberty and sense of moral responsibility."[50]

The primary organizer of the conference, the seminary's then president, Rabbi Louis Finkelstein, began corresponding with Urey at the end of 1939 and soon convinced him to join as one of its eighty-four founding members. Others among the conference's founders were Brooks, Robert M. Hutchins, Albert Einstein, I. I. Rabi, Arthur H. Compton, and Harlow Shapley.[51] Urey did not participate much in the planning or organizing of the conference's initial 1940 meeting, but once he was made a member of the conference's executive committee he began to take a more active role in the conference's activities.[52] Although Urey warned the committee early in 1941 that he and his fellow physical scientists were becoming increasingly preoccupied with national defense, he nonetheless agreed to preside over a public assembly at which Assistant Secretary of State A. A. Berle Jr. presented an address about the wastefulness of the Nazi forces, and to participate as a discussant during a session on "The Natural and Social Sciences in Their Relation to the Democratic Way of Life."[53]

Urey did not in fact manage to contribute much to the conference before war work consumed his time completely, and he failed to keep up with the conference's activities after the war. Still, he approached the conference seriously and seems to have been convinced of the importance of protecting the Judeo-Christian tradition in America. For more than a year, Finkelstein and Urey met and lunched on their common ground of Morningside Heights. Although no records of their personal discussions exist, their correspondence from this period displays mutual respect and agreement over the importance of the conference and its mission. When it came time for Urey to participate in the second meeting, he took his role as a discussant quite seriously. The unpublished notes from the Natural and Social Sciences session show that Urey had a great deal to say. It was in this session that Urey perhaps first articulated those ideas that would become so prominent in his 1950s speeches.

During the morning session he suggested that schools should do more to teach students the "Hebraic-Christian tradition that conditions all of our acts at the present time," instead of focusing exclusively on "Greek civilization":

> It is this religion that states that man is an important individual, regardless of his position in life, the color of his skin, or anything else that you wish to name, and it is that dignity of man which is established, I believe, as a result of this Hebraic-Christian tradition that is the most important thing in our democratic ideals today. It does seem to me that it would be well if this Conference could make a statement to the effect that more education in the field of the literature of our important religions in our common schools would be well worth-while. It would be my hope that such a statement could be made by this Conference before it closes.[54]

Urey even went so far as to wonder if science itself might be to blame for the erosion of the "permanent values" of Western civilization, saying, "I think it is time for us to consider the question as to whether science, with its materialistic point of view, may not be getting too strong a hold upon our ideals."[55] Margaret Mead, also present in the session, agreed with Urey that scientists must be made more aware of the social implications of their work, and stated that "if we look at the history of science, we find that the men who have meant most in the development of that science have been men who have been alive to the moral, religious, political implications of the points of view that they were helping to develop."[56]

Urey's views did not go unchallenged. As Fred Buettler argued in his thesis devoted to the conference's early years, its founders meant to address American pluralism by creating an inclusive dialogue between Protestants, Catholics, Jews, and secularists;[57] however, its discussions also included non-Western traditions like Buddhism and Hinduism. Howard University's Haridas T. Muzumdar, Mahatma Gandhi's spokesman in America, questioned Urey's neglect of "the largest segment of the religious and spiritual experience of the human race," and asked if imposing the Judeo-Christian tradition on the rest of the world was not a form of "spiritual fascism or totalitarianism."[58] Urey acknowledged that he had been speaking from his own limited experience, and softened his position on the teaching of the Judeo-Christian religions in public schools:

> In my proposal this morning, I was quite partisan. . . . I am looking about for a way to insure [our way of life's] continuance at the present time and, in so doing, have attempted to assess what the fundamental thing is that has given us this life of which we are so fond. I believe it is a Christian way of life and that Christian way of life is the mother of our democratic way of life. That is the way I personally look at it. . . . I propose that we study those religious beliefs, as literature in our schools, as a way to the definite end of combating totalitarianism in the world at large.[59]

Still, the majority of Urey's remarks that day remained committed to the role that Judeo-Christianity had played in the development of Western civilization.

Toward the end of the day, Urey objected to the thesis of the psychologist Max Schoen that freedom, "this right of every man to his own life, on his own terms, is not merely a proclaimed right, but one that is deeply rooted in the very nature of animal existence in general and of human existence in particular," and that therefore democracy is "neither a wish nor a hope, but the only mode of communal life in which there can be peace and which can have permanence."[60] Here Urey claimed that democracy is not something inherent to the human psyche, but something that historically had to be foisted upon the "savage" human:

> I would say it is surprising to me that such a large fraction of past history has consisted largely of despotism; in fact, history to me looks more like despotism with small interludes of democracy, rather than what I should expect upon the basis of this thesis, namely, universal democracy with an occasional despotism. . . . I think it is unscientific not to recognize facts from observation, and I believe that the statement Dr. Schoen presented fails to recognize facts. Most of us who are not descendants from the Mediterranean region are descended from savages as recently as 1,500 years, savages who left nothing but the rudest stones as monuments. Superimposed on that background has been a past adoption of civilization from other re-

> gions, and those regions are Judea, Greece, and Rome. I think it is well to remember that it is this sort of tradition that has led to our democratic institutions and no reasoning about our biological characteristics at all, even if they are true; and that it is this past tradition that has given us ideals that lift us above the level of the common, ordinary, primate.[61]

When the *New York Times* reported on the day's discussions, they picked up on Urey's contention that "it was no accident, that the totalitarian States were 'anti-Christian,'" as well as his claim that the Judeo-Christian tradition was "the mother of democracy."[62] Either of these claims could easily have found a home in a Brethren publication from the turn of the century. Had Urey had a chance to continue his involvement with the conference, he might have been able to refine his views over the course of the next several years as their annual meetings continued. But war was on the horizon, and Urey was to be swept up into one of the most stressful work environments he would ever experience.

ISOTOPE SEPARATION AND WORLD WAR II

The world was on the verge of becoming yet more volatile. In early 1939, as tensions were rising in Europe and Germany was preparing its invasion of Poland, the Jewish physicists Lise Meitner and Otto Frisch—in exile in Stockholm and Copenhagen—reported their discovery of a new type of nuclear reaction that they termed nuclear fission. Late in 1938, Otto Hahn and Fritz Strassmann in Berlin had bombarded the heavy element uranium with neutrons and split the uranium nucleus.[63] The resulting emission of neutrons from the fission of uranium suggested to the physicists the potential for a powerfully explosive chain reaction. As conditions in Europe moved closer to military conflict, this announcement led to great concern among physicists in England and America. Particularly excited were émigrés such as Enrico Fermi, Leo Szilard, and Albert Einstein, who had fled the Nazis and Italian Fascists in Europe. They worried that this information and its implications were also obvious to the Germans. A movement grew among the physicists at Columbia, Princeton, and the University of Chicago to impress upon American political leadership the destructive potential of nuclear fission and the importance of developing atomic capabilities before the Germans.[64] With belief in the superiority of German science still pervasive, the urgency of this message was understandable.

In March 1940, the Columbia physicists Eugene Booth and Aristid von Grosse, with the engineer John Dunning, confirmed that U-235 was the fissionable uranium isotope using a small sample concentrated in University of Minnesota physicist Alfred O. Nier's mass spectrometer. By this time, Urey had already begun considering methods for separating the isotopes of heavier elements—proposing in print a method that would utilize a countercurrent flow centrifuge. He was also by now the recognized world leader in isotope separation. In April 1940, Urey joined a group of concerned Columbia faculty members. This group, headed by George Pegram, a professor of physics and dean of graduate studies, proposed to the Naval Research Laboratory and President Roosevelt's Committee on Uranium that research on the separation of uranium isotopes should begin at once at Columbia. The US Navy in return asked Urey to organize an advisory committee of experts to counsel the Committee on Uranium. This group, which included Urey, reviewed the uranium problem at the Bureau of Standards in June 1939, and recommended immediate investigations of both isotope separation and the chain reaction.[65] That fall, Urey began work on isotope separation by centrifuge under a navy contract.[66]

Urey was eager to get American scientists involved in the war effort even before the perceived atomic threat. After the outbreak of the war, he distributed lapel buttons to his students and assistants that read "Defend America by Aiding the Allies."[67] But Urey's preoccupation with the deteriorating situation in Europe had begun much earlier. Kept up to date on events there through the reports of his European friends and colleagues, he feared for their safety and commiserated with them over the rise of fascism.[68] When the situation threatened his colleagues, Urey did what he could to secure positions for them in the United States, as well as to help them flee. Most famously, when Mussolini decreed anti-Semitic law in Italy to appease the Nazis in 1938, Urey helped Enrico Fermi, whose wife, Laura, was Jewish, flee Italy and make a home in Leonia, New Jersey.[69]

Urey's concern went beyond the Fermis. In his letters to de Hevesy during this period, Urey reported his desire to help the Jewish physicist in whatever way he could, and also related his frustrations at the "very strong feeling in this country for keeping out of the war."[70] Urey helped de Hevesy open a joint bank account in New York City and deposited $1,000 of de Hevesy's money so that he could bring his family to the United States if he had the chance. After the invasion of Denmark, Urey worked with Warren Weaver of the Rockefeller Foundation to attempt to locate de Hevesy and get him out of Europe on a three-month invitation from the foundation.[71] A letter to Linus Pauling, in which Urey explained that Otto Redlich had been fired from an Austrian university for being Jewish and asked whether or not Caltech could find a position for him, shows that Urey had become involved in helping Jewish scientists flee Europe as early as 1938.[72]

It is not surprising, then, that Urey took Fermi and Szilard's warnings seriously and was one of the earliest supporters of work on nuclear fission at Columbia. He accepted without reluctance the position of chairman of the Advisory Committee on Nuclear Research that would give technical advice to the president's Committee on Uranium. Once work had begun, Urey coordinated investigations into possible means of uranium isotope separation at the University of Virginia, Harvard, and Columbia, where teams experimented with centrifuges, gaseous diffusion, and chemical separation.[73]

While Urey was eager to contribute to the war effort, he did not enjoy the work of researching and designing an industrial process for isotope separation. Although he was skilled at building and tinkering with instruments in the lab, like many scientists he was not well suited to managing a large project. Where he felt he could be most useful was in producing heavy water for reactors. However, the engineer John Dunning was also working on uranium isotope separation methods at Columbia, and the army wanted only one director for all of Columbia's work. Because Urey was a known expert on both heavy water and isotope separation, he was a natural choice to lead both efforts. "I had been trying to separate isotopes by chemical means," he told Groueff. "It was a tough job. The theory was not too difficult. I had worked out the theory of this. The carrying it out was a chemical engineering job, again. I was thoroughly tired of it by the time the war started, and then I had to keep this up during the war."[74] Urey also later complained that the program grew too large and cumbersome for him to handle. From mid-1940 to mid-1941, Urey was working with a Columbia staff of five faculty members, three other personnel, and a research budget of $29,700. At the end of 1941, the program moved from the research stage to the engineering and construction phases. By the end of 1942, Urey had a staff of 180 technicians; by the end of 1943, Urey had more than seven hundred people working on gaseous diffusion alone, with hundreds more at universities and laboratories throughout the eastern United States.[75]

Urey later described his position during the war as "a sort of glorified personnel officer" and complained, "I didn't do any scientific work myself."[76] As his research assistant from this period, Karl Cohen, remembered, "He had little taste for administration, and the burden weighed heavily on him."[77] Urey's unhappiness was also felt at labs outside of Columbia. James Arnold, who worked on isotope separation "about three tiers below" Urey at Princeton, said, "The research project at Columbia became very large, and deeply involved with engineering. My professors at Princeton thought this aspect was uncongenial for Urey. This view is consistent with the character of the man I knew later."[78] And Urey himself would later tell an interview that the war, besides heightening his own dislike for the atmosphere of Columbia, also added new dimensions of unpleasantness to his life in New York:

> I was most unhappy during the war. I had bosses in Washington who didn't like me, and I had people working for me who didn't like me. Imagine a more miserable situation—where you can't resign, but nobody wants you around! About the worst situation you can get in. When the war was over I got out. I was very close to a nervous breakdown during the war. Old General [Leslie] Groves would send his physician around to look me over. . . . After the war he saw how perked up I was and so forth, he wondered what had happened to me. "Well," I said, "I have good bosses, that's all."[79]

In another interview, Urey admitted that he felt General Groves was one such boss who didn't like him, and claimed that Groves had been "very suspicious" of him during the war.[80]

Urey was correct. Above him, Groves had little patience for Urey and developed a low opinion of the distinguished chemist. From Groves's first visit to Columbia, he was skeptical of Urey's ability to manage the project: "I was not particularly impressed with Urey because he seemed so uncertain in his answers to questions and in his general grasp of the project." Groves also noticed that Urey did not get along with Dunning, who was doing most of the technical work on the project. "It was obvious that Dr. Dunning and Dr. Urey were at outs," Groves later recalled. "Dunning had no respect for Urey. Urey thought that Dunning, not being a chemist and not being a Nobel Prize winner, he couldn't amount to much either. There was definite animosity." This added up to what Groves described as an "impossible situation" stemming from the "inability of Urey to organize his laboratory and the animosity of Dunning and the contempt that Dunning had for Urey."[81]

Dunning and his group resented Urey for taking charge of what they felt was their territory. Urey was convinced that Dunning was "very jealous of his ideas in regard to separating uranium isotopes by diffusion," and that he did not want Urey interfering or attempting to supervise his work. "It was my impression that he regarded it as his baby," Urey told Groueff. The two did not grow any closer over the course of the war. As Urey saw it, Dunning was young and interested only in self-promotion. "He did not want the famous man around. That is all."[82] Dunning was a young man but he had already developed a reputation as a political maneuverer. Groves recognized that Dunning was a "go-getter type" and had a stronger personality than either Urey or Pegram; he walked all over them with his "great talkativeness" and "domineering personality," as well as his exuberance and confidence. In his attitude, Groves found him to be the exact opposite of Urey: "He was such a great optimist that you couldn't believe what he said."[83] The two men's relationship grew toxic. Urey eventually became paranoid about Dunning's intentions. He told Groves's chief of staff, Kenneth Nichols, "I think John Dunning just backs the gaseous diffusions plant because he is out to ruin my reputation. He wants me to be the leader of a failure."[84]

Another part of Urey's frustration came from the several reorganizations of the program during the war. These ultimately, and necessarily, replaced "the Greek democracy of volunteer scientists" with "central direction and mission-oriented laboratories."[85] The first such shift reorganized the Science Advisory Committee to the Committee on Uranium and placed it under the National Defense Research Committee (NDRC), headed by Vannevar Bush.[86] As a member of a new Committee on Uranium (which now omitted foreign-born scientists), Urey was given broad responsibilities for formulating the entire research program in isotope separation, and given a budget of $100,000 for the task.[87] But this arrangement too would soon be reorganized. The creation of the Office of Scientific Research and Development (OSRD) under the Executive Office of the President saw Vannevar Bush replaced by Harvard University's James B. Conant as the head of the NDRC; Bush became the head of the OSRD. The Committee on Uranium became the S-1 Committee of the OSRD, and its membership was expanded.[88]

Urey remained a member of the committee, although Bush and Conant now assumed much of his authority for the overall program in isotope separation.[89] Urey did not respond well to this new arrangement, and was particularly resistant to Conant as a supervisor. He would later deride Conant as a poor manager. During the war, Conant suspected Urey of expressing his dissatisfaction with Conant's management to other project scientists—an offense that he regarded as disloyalty. According to Conant's son, his father had "very little use" for Urey, whom he believed "shot from the hip and acted emotionally."[90]

But Urey's frustrations also had to do with the pressures of bringing a facility online for the separation of uranium isotopes. By the end of 1942, the S-1 Committee had settled on a gaseous diffusion plant, and had been granted a budget of $100 million for the plant's construction in Oak Ridge, Tennessee. The plant would be designed and built in collaboration with the Kellex Corporation, whose engineers promised that a six-hundred-stage pilot plant could be built within ten months. Urey, however, was not convinced that all the pumps, seals, instruments, controls, valves, pipe assemblies, and barriers could be designed and built within this time period—and it was he who bore the responsibility of overseeing these developments.[91]

The six hundred barriers through which the uranium gas would be diffused were one of Urey's biggest headaches, and were also the source of tension between Urey and Groves. Two barrier designs—one that did not work well but that Urey thought could be improved and another that was untested and would cause a holdup in plant construction—brought Urey and Groves into conflict. As Cohen remembered, "With ten thousand workers building a huge diffusion plant at Oak Ridge, Groves had to have a successful barrier, even if somewhat late. Urey was not prepared to redefine his objectives to call a late plant a success."[92] Urey was worried that a delay in the construction of the pilot plant would make it unlikely that the plant would play any relevant role in the war effort.

Hugh Taylor, a Princeton chemist working with Urey on the heavy water problem in Trail, British Columbia, remembered the toll that the barrier work took on Urey: "Urey was an extremely harassed man. . . .

The barrier job was the toughest job in the whole Manhattan Project. . . . It really looked hopeless. You had to have sublime faith to be encouraged to go on. And Urey was so nervous and excited about the whole thing."[93] Groves agreed: "He was extremely nervous. He was so nervous that in talking to him at lunch one day, he was unable to take a glass of water to his mouth with one hand. He had to prop both elbows on the table, place the right hand with his left hand on his wrist and then raise the glass to [his mouth] and it was still shaking." Groves also remembered that Urey "was scared to death of me. If I looked at him, he'd start to quail."[94]

In her published memoir of these years, Laura Fermi described seeing Urey on a train to "Site Y," the wartime code name for the atomic facilities in Los Alamos, New Mexico:

> It would be more accurate to say that through the open door of a roomette I saw a tired-looking man who looked like Harold Urey, stretched on the divan, absorbed in who knows what thoughts and what deep concern. . . . Harold was overworked and tired and looking older than his age all during the war years; he recovered only when he could put his mind at rest about the war and his wartime duties.[95]

The thought of Samuel Urey's mental exhaustion, breakdown, and institutionalization must have occurred to Harold during this time.

Groves and his subordinates attempted to deal with the problem by removing Urey as much as they could from the day-to-day operations of Columbia's code-named Substitute Alloy Materials Laboratory (SAM Lab): "We tried to get Urey interested in other things. We sent him to England on a trip to look at certain things over there hoping to see what they were doing, but essentially to get Urey away." Trips like these were failures, as Urey would return to Columbia at the earliest date possible. In early 1943, Groves decided to send Urey away on one last trip. This time he sent him to British Columbia, ostensibly to inspect operations at the Manhattan Project's newly upgraded heavy water plant in Trail. In reality, the trip was designed to be a forced holiday, including a weeklong fishing trip that might give him a chance to rest. Urey resisted. "He only stayed about twenty-four hours," Groves lamented, "and then he was just so keyed up that he couldn't stay." Hugh Taylor was put in charge of Urey during his visit, and recalled that the army told him, "Will you please keep hold of Urey for the next week. He's on the ragged edge. Take him out into the woods and let him have a holiday." The army's request came to naught: "[We] took him out into the woods on a Saturday afternoon. And we stayed for two days. And at the end of two days, he was itching to get back and we had to bring him out of the woods and put him on a plane and send him back."[96]

"You couldn't calm him down; you couldn't get rid of him," remembered Groves. So he found a way to minimize Urey's impact on the project. When Urey returned to Columbia, he was told that he had a new subordinate, Lauchlin M. Currie, a chemical engineer and vice president of one of the Manhattan Project's industrial contractors. Although Groves had given Currie the title of associate director, meaning that he was technically working under Urey, in fact he had been brought in to take over for him. According to Nichols, Dean Pegram explained to Urey, "Now, you can either be relieved of your responsibilities or you can sit in that office and issue no orders." Urey later admitted to many of his colleagues on the project that he agreed with this decision. "He told me once that I saved his life simply because he said he'd have worried himself to death," said Currie. When Nichols saw Urey on a train to Chicago after Pegram relieved him, Urey told him, "You know, Nichols, I'm starting to sleep again. I'm glad this has taken place."[97]

After Currie's arrival through to the end of the war, Urey remained the nominal head of the diffusion project. He remained at Columbia and worked with Currie, under whose leadership Dunning and his team fell in line. Urey's contribution to the project, in the end, was minimal. But Currie appreciated having the eminent chemist on the project for the purpose of bringing in scientists whom they otherwise would not have been able to attract. Urey tried to be as helpful as he could to Currie, but his enthusiasm for the project was long gone. The project eventually succeeded, after being handed off to the Kellex Corporation. The K-25 separation plant was, according to Nichols, "one of the greatest engineering achievements in history" despite all the troubles of the research and development phase.[98]

Years later, when working with the children's authors Alvin and Virginia Silverstein on a short biography of his life in science, Urey attempted to rewrite this history. He claimed that he had tried to pass the job of directing the Columbia team off onto Dunning from the start. In a dramatic reversal, he said that he recognized Dunning's skills as an organizer and administrator, and that he had more of the engineering experience necessary to do the separation work. The army gave him the job over these objections. This was a calculated plot, he believed, to pass off responsibility if the effort failed. As his view was represented in the Silversteins' biography, "the government probably had very little faith that the atom bomb project would ever succeed, and they wanted a Nobel Prize winner around to take the blame if it failed."[99] In this version of events, when he was removed as director it was not because of his failure to organize the SAM Lab, but because the army now saw that the project was going to succeed. He told Groueff something similar, concluding, "If the whole project had failed, it would be mine completely, but that if it would succeed then it would be somebody else's success. The latter has certainly proved to be true in the years since."[100] This attempt to change the story, and to make it a part of his official published biography, was clearly an effort to save face.

Thus while other scientists involved in the Manhattan Project would later admit to losing enthusiasm for the project after V-E Day, Urey's disenchantment began at least a year earlier. Also, as opposed to many of his colleagues, Urey not only came to regret unleashing the atomic bomb upon the world, but also developed an aversion to the isotope separation work that had built him into an eminent man of science. Urey later told Zuckerman, "[At] the beginning of the war I had been working on separating isotopes and I was tired of the job and at the end of the war I was still more tired."[101] Even more significant, Urey seems to have feared that increased support for that work would bring undue pressure: "[If] you tried to separate isotopes and it proved to be important to the Atomic Energy Commission to put the amount of manpower on the job, I couldn't possibly repeat that, you see."[102] Thus, in the years that followed the war, Urey would develop relationships with emergent government and military funding agencies while at the same time working to maintain a level of support that allowed him to direct a successful research program without becoming a manager.

CHAPTER FIVE

A Separation Man No More

> The unprecedented galaxy of scientists assembled there [for the Manhattan Project] began to disperse. They were anxious to return to their customary academic habitat, but with a new attitude: the wartime effort had ushered in "Big Physics," the use of large-scale equipment and the availability of massive financial support. Fermi, together with a group of other brilliant senior scientists . . . accepted offers from the University of Chicago. Some kind of "package deal" was involved (it is rumored that the same "package" had proposed themselves earlier to the University of Washington, but that the deal fell through).
>
> VALENTINE L. TELEGDI, "ENRICO FERMI, 1901–1954"

There is substance to the rumor hinted at by V. L. Telegdi in the above passage. Prior to accepting any new positions after the war (and, for that matter, prior to the use of the atomic bomb), the atomic scientists did in fact go shopping for an institution that they could shape into a center for postwar nuclear science, and where they could continue to work at the pace and scale to which they had become accustomed during the war. Urey discussed the prospect of moving as a group with his Columbia colleagues Enrico Fermi and Joe and Maria Mayer, as well as the physicist Edward Teller.[1] The group initially set their sights on the Pacific Northwest, where living conditions would be "more delightful" than what they were used to on the East Coast and in the Midwest. (Fermi was especially interested in enjoying "a more congenial place than the crowded cities of the east.")[2] On the group's behalf, Urey took a trip to the West Coast in summer 1945 and attempted to broker a deal.

In July, Urey reported to Teller that he had been to the University of Washington and had met with the president and the members of the chemistry department. While Urey reported favorably about the encouragement he received at these meetings, he ultimately decided that the prospects were not as good as they might appear. There were definite risks associated with a state university that had not yet been initiated into the wartime world of contract research. Urey explained to Teller that efforts at a public university would require the support of the board of regents and the state legislature, which would inevitably entail "a considerable educational program," and possibly "a number of years of struggle without any certainty that we shall succeed."[3] Urey told Fermi around the same time that no one at Washington seemed to have any idea of how to organize or fund such an institute.[4] Furthermore, Urey was able to convince Washington to make definite offers just to Fermi and himself. He was not sure he could get offers for Teller or the Mayers, and he was determined to keep the group together.

The prospects at Chicago, on the other hand, looked much brighter—so bright, in fact, that "the Washington plans faded rapidly in[to] the background."[5] Prior to his trip west, Urey felt he had been receiving cautious hints about a postwar position at the University of Chicago. Chicago had, like Columbia, contributed to the Manhattan Project. The physicist Arthur Compton had directed efforts at the university to extract plutonium from chain-reacting "piles" of uranium, and to develop a weapon from the fissionable material. The effort had brought together scientists from the East and West Coasts, in a project that was code-named the "Metallurgical Laboratory" (Met Lab). Urey knew that Compton had already approached the university's president, Robert M. Hutchins, about finding a way to keep the roster of talented physicists, chemists, and engineers under one roof after the war. Hutchins was receptive and came to the conclusion that he would be able to raise the necessary money for a new institute only if he had a stellar group of scientists with reputations like Urey's to help found it.[6]

Upon his return to New York through Chicago, Urey received a firm offer. The university would found a new Institute for Nuclear Studies. In addition to establishing a position for Urey at the new institute and in the Chemistry Department, Chicago told him, they had offered a position in the Physics Department to Teller, and they were seriously considering Joe Mayer. Sexism in the elite schools was persistent; Chicago ultimately would offer Maria Goeppert Mayer an unpaid voluntary associate professor position—and yet she would later win the Nobel Prize for her work on the nuclear shell structure, done while at Chicago.[7]

Chicago also informed Urey that Cyril Smith had already accepted a position and that they were continuing to try and persuade Fermi to join. "It therefore seems to me that the group that we were thinking of will all appear together at Chicago," Urey wrote to Teller.[8] Equally important was the guarantee on Chicago's part that "plans were under way for adequate funding from private sources to finance a big development."[9] This showed that Chicago "understood the trend of the times and that it would not confine the activities of basic research to the meager laboratories and still more inadequate funds available before the war."[10] Hutchins and the scientists agreed that the institute would be "a meeting ground for science and industry"; industry would fund the institute in exchange for scientific advice and access to cutting-edge research.[11]

GREENER PASTURES

In addition to his desire to remain a member of the core group of atomic scientists in their new institutional home, Urey was uncomfortable at Columbia University. Even though he had done his Nobel Prize–winning work there, and had led a successful interwar research program in isotope separation, he had never felt at home. Even before the stress of supervising Columbia's uranium separation work, and before the prospect of a move to Chicago, Urey had been looking for an escape from New York City. Now, at the end of the war, the basement lab of Pupin Hall, where Urey had first modified his grating spectrograph and detected deuterium, had been all but taken over by his wartime nemesis, John R. Dunning, and his cyclotron.[12]

Chicago offered Urey a landing place where he could get a fresh start away from the pressures of war and the stress of interpersonal conflict. However, before he accepted the offer, he made sure that he would be accompanied by those colleagues with whom he felt a mutual respect and fondness. Throughout summer 1945, Urey conferred with members of the Columbia "package." Meanwhile, negotiations continued between

Urey and Chicago, now represented by Robert S. Mulliken (a fellow isotope chemist who had served as the director of the Information Office of Chicago's Plutonium Project). In August 1945, Mulliken and Urey sat down to discuss future research programs at Chicago.[13] After this, Urey summoned Maria and Joe Mayer to Chicago for a three-day conference concerning the institute and asked that the three of them discuss their collective position in person beforehand.[14] Meanwhile, Urey also drew up a plan for an isotope separation program at Chicago that required roughly $100,000 for salaries and equally as much for instruments, and sent this plan to Chicago's dean of physical sciences, Walter Bartky.[15]

To Urey, who felt equally at home with physicists as well as chemists, it was important that the institute continue the wartime trend of ignoring any boundary between the two disciplines. By December, the physicist Samuel K. Allison, who would be the institute's first director (Urey and Fermi both declined the position), was looking outside of Chicago's Met Lab for recruits. Allison later described the plans for the institute: "It is the avowed purpose of the Institute to have physics and chemistry under the same roof." All the senior members of the institute were to have a joint appointment in the institute and either the physics or chemistry department, and would instruct courses and advise graduate students within those departments; the institute itself would not be a degree-granting entity but a place for cutting-edge research.[16]

In a 1947 article in *Scientific Monthly*, Allison went further in his description of the institute's marriage of physics and chemistry, this time in more nuclear terms: "In it an attempt is being made to attain the complete fusion of the sciences of physics and chemistry, at least in their advanced aspects."[17] He defined the main three tasks of the institute as researches in nuclear physics, radiochemistry, and chemistry of the separation of isotopes. But despite these foci, the lab was not organized around specific research programs. Instead, "Each member [was] free to follow what seem to him interesting and promising investigations."[18] One wartime trend that continued at the institute was an in-house staff of engineers and technicians. Unlike other departments within Chicago's Division of Physical Sciences, the institute had the authority to make both academic and technological appointments. Allison explained that the institute was an "outgrowth of the experiences during the war, in which physical scientists from many fields cooperated with engineers and technologists in the successful effort to liberate nuclear energy in macroscopic amounts."[19] Indeed, in 1953 the institute's "approximately fifty electrical and mechanical engineers, draftsmen, and research assistants" more than matched its thirty-eight faculty members and research associates.[20]

MOVING AWAY FROM ISOTOPE SEPARATION

When Urey first negotiated with Chicago for his postwar position, no atomic weapons had yet been used in war. The isotope separation program that Urey outlined to Bartky was postmarked from New York on the very day of the Hiroshima bombing. By the time the war had ended, Urey had been so "traumatized" by his experience as director of Columbia University's SAM Lab that he could no longer muster any enthusiasm for the prospect of continuing isotope separation work.[21] Urey's colleagues had witnessed his breakdown, although in describing it they never drew attention to the personal conflicts between the chemist and Dunning, well known as these may have been. Hutchins claimed to have witnessed Urey becoming more and more "disturbed" as the project "drew closer and closer to what I now regard as its catastrophic end."[22] According to Joe Mayer, the trauma of war work stuck with Urey for some time even after taking up residence at the institute, causing him "to drift, looking for new fields to conquer."[23] And Urey's collaborator Hans Suess remembered that, while most scientists were able and eager to return to their prewar research programs, Urey was "anxious to get away as far as possible, in time as well as in space, from everything connected with weaponry and means of destruction," including his prewar work on isotope separation.[24]

Aimlessness and angst were not characteristic of Urey, who before the war approached his scientific projects with great enthusiasm and what his colleagues described as a childlike curiosity.[25] In his Nobel address more than a decade earlier, he had excitedly reported the thermodynamic properties of isotopes to the world and had spent a considerable part of the address speculating about the possible methods of separating the isotopes based on these differences—work that, once put into practice at Columbia with a string of graduate students, lab assistants, and grants from private foundations such as the Carnegie Institution, came to define the research program in Urey's lab up through the war. This work was mainly of interest to a relatively small cohort of physicists and chemists researching the structure and behavior of the elements and their isotopes, and to an even smaller group of biologists eager to use these isotopes as experimental tracers. During the war, however, the increased pace and pressure of industrial-scale isotope separation had nearly broken him.

If Urey's war trauma was the primary reason for his aimlessness in the immediate postwar years, his work for the control of atomic weapons was a close second. Urey's activities with the scientists' movement consumed him for the first few years after the war. He told the *New Yorker*, "I've dropped everything to try to carry the message of the bomb's power to the people . . . because if we can't control this thing, there won't be any science worthy of the name in the future."[26] This was a great commitment. Even though Urey had been a popular public speaker before the war, his new pace, combined with the urgency of the atomic problem, was difficult for him to handle. "Publicity ruins a scientist," Urey said. "The phone rings all the time and you can't settle down. I'm thinking of ripping out the phone and changing my name. I have a stack of mail I haven't even read, let alone answered. Hell, I'm no public figure! Who am I not to be reading my mail?"[27]

REORDERING THE WORLD FOR THE ATOMIC BOMB ERA

> Among my scientific colleagues few have devoted themselves wholeheartedly to the cause of enlightenment as Professor Harold Urey. He has shunned no sacrifice of time and energy when it came to serving our important aim. Whoever is himself filled by the passion for scientific research knows how hard it is for a man of our kind to abandon his own aims for a length of time and to serve a social task, simply out of a feeling of duty and of necessity. I may say in the name of all of us that we are thankful to him for his untiring unrelenting efforts and we hope that his words which are based on sound knowledge and on a feeling of responsibility will find fertile soil.
>
> ALBERT EINSTEIN, RADIO ADDRESS, NOVEMBER 17, 1946

Even after the introduction of atomic warfare, Urey's optimism for world improvement initially remained strong. Through organizations like the Atomic Scientists of Chicago, Urey and his colleagues attempted to consolidate the opinion of their fellow scientists on their role and responsibilities concerning atomic power so that they could present a united front before Congress in the hope of influencing atomic policy. By November 1945, they had formed a lobby named the Federation of Atomic Scientists (later changed to the Federation of American Scientists and referred to as the FAS). In its first year, the federation attracted 2,500 dues-paying members, established offices in Washington, DC, and hired a press agent.[28] Meanwhile, with his scientific hero, Albert Einstein, Urey helped found the Emergency Committee of Atomic Scientists. The Emergency Committee, composed of eminent members, was formed to assist the FAS in fundraising activities for the National Committee on Atomic Information, which had as its primary goal public education on atomic energy and its societal implications. Concerned over too casual an acceptance of atomic weapons, the atomic scientists hoped that a properly educated public would fear the bomb's destructive capabilities as the scientists did. Thus, within the first year after the dropping of the bomb, they had developed a two-pronged approach that focused efforts on public education and political lobbying.[29]

At the close of 1945, the *New Yorker* reported on the educational activities of the atomic scientists, and pronounced that seeing the scientists "pop out of their cloisters all over the place in order to issue warnings" about the bomb was evidence that the social responsibility of the scientists was indeed "well developed."[30] As a charismatic and articulate member of the committee, Urey played his part. He made multiple visits to Washington, DC, where he testified before the Senate Special Committee on Atomic Energy. While Einstein wrote letters from Princeton asking for donations on behalf of the Emergency Committee, Urey toured the country addressing various interested and influential organizations. The "heavy-water man and Nobel Prize winner" told one audience, "I know the bomb can destroy everything we hold valuable and I get a sense of fear that disturbs me in my work. I feel better if I try to do something about it."[31] When asked if he felt guilty for having unleashed the bomb upon the world, Urey responded that guilt didn't enter into it. While he regretted helping to weaponize atomic energy, "atomic energy is in nature. . . . It can't be concealed. Scientists can't prevent modern war by refusing to do scientific work. The solution is political."[32] Urey was convinced that the possible peacetime uses of atomic energy would eventually eclipse the threat of the bomb, but only if adequate controls could be placed on weapons. He was determined to see such measures imposed before returning to his own research program, and he hoped to see this happen soon.

Toeing the Emergency Committee's party line in a *Collier's* article that the committee reproduced and distributed, Urey told the world that he was frightened of the bomb: "I write this to frighten you. I'm a frightened man, myself. All the scientists I know are frightened—frightened for their lives—and frightened for *your* life."[33] In place of Urey's prewar descriptions of science as a search for truth and beauty, humankind's potential for good and evil had taken up residence. While Urey in his earlier speeches had cautioned that science might one day through man's belligerence destroy Western civilization, this potentiality now took on the character of an impending apocalypse and moved into the foreground of his rhetoric. The invention of the atomic bomb had put humankind "face to face with the powers which, philosophically speaking, are supreme in our universe," and unleashed once and for all the "limitless powers of the universe as developed by the limitless imagination of Man."[34] This had accomplished nothing less than to reorder time and inaugurate a new calendar era: "This is indeed The Year Atom Bomb One," Urey wrote. "It has opened most ominously. We must waste no time if we plan to be alive in A.B. 5 or A.B. 10. Atomic war could unleash forces of evil so strong no power of good could stop them. The main race, between man's powers for evil and his powers for good—that race is close to a decision. You must think fast. You must think straight."[35]

When asked how far back in history one must search before finding a discovery as significant as that of atomic energy, Urey argued that the closest thing he could imagine was the discovery of fire: "If one thinks then that we have a discovery, made in the last few years, for which we cannot find an analogy except by going to prehistoric times, we must expect that we have before us a new source of energy that is likely to create an emergency that will last much more than a matter of months or a year. It will be an emergency that will continue for many years in the future." Just as fire had helped fundamentally to reshape the lives and social structures of prehistoric humans, Urey was certain that atomic energy too would reshape modern society.[36]

For Urey, and for many of the scientists involved, atomic weapons demanded that the world be reorganized.[37] His prewar speculation that the many nations of the world might inevitably become a "world state" by the year 2000 now became a plea for world government.[38] With this aim in mind, Urey, Einstein, Edward Teller, Harrison Brown, and Leo Szilard all personally endorsed the United World Federalists. The prospect of world government was particularly attractive at the University of Chicago, where in 1946 Hutchins organized a committee of professors to debate and draft a "World Constitution."[39] After two years the committee produced a draft that began with the preamble, "The age of nations must end, and the era of humanity begin."[40]

Urey found a precedent for world government within the scientific community. He told his *Collier's* readers that he knew scientists from every corner of the world, and that he had learned before the war that they all spoke the same language. Now that the atomic bomb had entered the world, the scientific community would become even closer, brought together as it was by "a common fear and a common pledge and a common hope." In this early plea for world governance, Urey even expressed confidence that the Russians would want to be involved. After all, Russia had her own native scientists, and they would carry the message of fear to their leadership: "If you realize, as scientists do, that Russian science includes some of the best brains in the world today, I think you will understand, first, that Russian leaders must naturally be frightened of the possibilities of this power and, secondly, that it will not be long before they also are masters of it." Urey argued that no country had been more devastated by the Second World War than the Soviet Union, and the Russians had come to understand this war's great consequences as had no other country: "No one who understands atomic war wants anything but peace." In a moment of naivete, Urey imagined a meeting of Russian and American leaders, along with their scientific advisers, in which the scientists acted as mediators between the two groups: "Scientists will have no trouble understanding one another. When they meet I think their recommendations will be almost unanimous."[41]

THE RUSSIAN PROBLEM AND THE ATLANTIC UNION

Urey's speeches from these immediate postwar years advocated strengthening the United Nations into a sovereign world government that had "adequate powers to prohibit atomic bombs" and "to police the world to see that such laws are obeyed."[42] The alternative, which Urey always claimed was the probable result of a lack of world government, was to cover the world in "armed camps"—outposts of American missiles matched by those of the Russians—and live "in constant fear that the other man's itchy trigger finger would start something moving in a very short time."[43] In addition to policing the world for atomic weapons, the world government would also help ensure that there was no war: "First of all, let us note that the atomic bomb is not the fundamental problem at all that we have to face. The fundamental problem is war. If there is another war, atomic bombs will be used."[44]

Urey's hope that the Soviets would participate as partners in the world governance of atomic weapons was short lived. In March 1947 the Soviet Union rejected the US-proposed Baruch Plan. The plan was based on the *Report on the International Control of Atomic Energy* (bylined Dean Acheson, secretary of state, and David Lilienthal, chairman of the Tennessee Valley Authority, but largely written by J. Robert Oppenheimer). The American statesman and financier Bernard Baruch presented it before a session of the United Nations Atomic Energy Commission on June 14, 1946.[45] The plan proposed an exchange of basic scientific information among all nations, control of nuclear power that would limit its use to peaceful purposes, elimination of atomic weapons and other weapons of mass destruction, and international inspections to be performed by an international atomic development authority, which would also oversee the mining and use of nuclear materials, and would grant licenses to countries pursuing peaceful nuclear research.

Urey had supported the Baruch Plan's establishment of an atomic development authority, writing to colleagues in the Atomic Scientists of Chicago that "Mr. Baruch's proposal was that the ultimate decision in regard to this matter of the atomic bomb shall be made by the Atomic Development Authority, and not by the nation-states, and that proposal is the only one that makes sense from the standpoint of real prevention of the use of any weapon. We are not afraid of the state of New Mexico, in spite of the fact that atomic bombs are made there, only because New Mexico has no power whatever to use those bombs."[46] Furthermore, Urey was happy that the plan would force the United States to dispose of its nuclear weapons: "Since I do not believe that atomic bombs should be made by anyone [and] none should be possessed by any government of any kind, I cannot look with favor upon any other country's learning how to make bombs, and I can only hope that arrangements can be made such that the United States will forget this art."[47]

The Soviet rejection of the Baruch Plan did not completely deter Urey from pursuing some form of world government. He became an ardent supporter of Clarence K. Streit's Atlantic Union, and favored a nuclear alliance that would strengthen the West against the Soviet threat, even at the high cost of an arms race. This led to a split among the atomic scientists over the best course of action after the Baruch Plan. The other side of the split was held by Urey's colleague and compatriot Leo Szilard, who argued that any union without Russia would lead to a nuclear arms race and war; Szilard held out hope for a worldwide reconstruction of economic, social, and political relations.[48] Urey stated in a radio interview in 1949, after the Soviets had demonstrated their own atomic capabilities, that he concluded only two months after the Baruch Plan was introduced that the USSR would not agree to any effective control measures: "I began immediately to revise my opinion as to what the proper course for the people of the United States should be."[49] The Soviet Union's refusal of the Baruch Plan had strengthened Urey's distrust of the Soviet leadership. He supported the Truman Doctrine and the Marshall Plan as steps toward an Atlantic Union:

> We do not intend to become a Soviet Socialist Republic and will accept atomic war first. We are determined to fight the Communist dictators of Russia in any way possible and in any part of the world. We as a people have adopted this view because of our observation of the behavior of the cruel and ruthless dictatorship of Russia, as we have observed it in operation since 1917, and particularly in light of discussions in the United Nations since the war. I further believe that we have adopted this view because we believe that the USSR has aggressive intentions toward her immediate neighbors. This has been abundantly confirmed since the war. We further believe that these aggressive intentions are probably not limited to the European countries. The difficulty between the USSR and the United States is partly a power conflict, it is true. But the power conflict is founded upon a profound difference in philosophy.[50]

This profound difference between the United States and the Soviet Union, now backed up by atomic weapons on both sides, led Urey to conclude that war between the two powers was more likely than ever—particularly if both powers felt that they could win such a war. "The only course of action which will enable the United States to avoid war," Urey argued, "is one which will make the West stronger. I have maintained since 1946 . . . that the most effective way to increase the strength of the West is through the formation of a federal union of the Atlantic democracies. I believe that the Truman Doctrine, the Marshall Plan, the Atlantic Pact [NATO]—all of which have been approved by Congress—are all steps leading in this same direction."[51]

Thus Urey eventually became a supporter of the development of the hydrogen bomb and of stockpiling atomic weapons. These measures, along with the adoption of the Atlantic Union, were the only way "the Western democratic powers [would] be able to maintain an overpowering political, commercial, military, and ideological strength. Only in this way can we have an enormous unbalance of power, so that perhaps one side does not attempt to start a war because it recognizes that it cannot win, and the other side does not need to start a war because it knows that the weaker side will not dare to attack."[52] As he explained in a letter years later, he had by then given up on a world constitution, believing that "those who engage in this sort of thing are completely unrealistic."[53]

THE COLD WAR AT HOME

In addition to the Soviet refusal to participate in world government, American political reactions to the Cold War presented a Sisyphean challenge for intellectuals who questioned excessive security-loyalty measures or isolationism. As early as 1946, US Congressman J. Parnell Thomas attacked Urey on the floor of the House of Representatives for opposing the May-Johnson bill. The bill would have extended the military control and clandestine character of nuclear research that had prevailed during wartime. Thomas, a member of the Military Activities Committee and ranking Republican member of the House Un-American Activities Committee (HUAC), attacked the McMahon bill Urey supported (which advocated civilian control of atomic research and formed the US Atomic Energy Commission) as being "the creature of impractical idealists." Thomas likewise attacked Urey, labeling him a "one-world-minded" person who was "blinded" by an "intense ardor for a better world." As

Jessica Wang has argued, such political attacks and the anxiety of secrecy and surveillance pushed many progressive left-wing scientists like Urey toward more cautious and conservative Cold War positions.[54]

Although the Federation of American Scientists succeeded in killing the May-Johnson bill, this and other political organizations of the atomic scientists were put on the defensive almost immediately. Their ideology of international cooperation and intellectual freedom clashed with what Wang has described as "the postwar preoccupation with national security and protection of the 'secret of the atom.'"[55] As the legal scholar Walter Gellhorn observed at the time, "The world's polarization into opposing forces has cast a shadow upon the traditionally accepted values of scientists. In days gone by science was broadly viewed as an unselfish effort, international in scope, to expand knowledge for the benefit of all mankind. Today science has come to be regarded somewhat in the nature of a national war plant in which a fortune has been invested."[56]

The world had changed, and world government advocates like Urey now found their loyalty to their country being questioned. By early 1947, the stress of working for world government while being attacked by HUAC had exhausted Urey. He wrote to Einstein that his doctors had ordered him to avoid outside activities: "I find that I am able to carry my university work and that is about all. Otherwise I become very tired, unable to sleep, and generally quite unable to take care of any of my work."[57] As one of the Emergency Committee's most active public speakers, Urey knew that he could not walk away from the committee without leaving it severely weakened. But he pleaded with Einstein to consider adding "other outstanding men," such as J. Robert Oppenheimer, who could add their prestige to the group while shouldering some of the burden.

As atomic scientists found their energies redirected toward the defense of their own civil liberties, their political lobbying efforts were severely hampered. The Emergency Committee spent much of 1948 rallying to the defense of the physicist and head of the Bureau of Standards Edward U. Condon, who was attacked viciously by HUAC. The committee organized a dinner in support of Condon that was meant to "indicate to other scientists that we, as the scientific fraternity, will stand together."[58] The dinner also expressed their disapproval of HUAC's activities. In his letters to other prominent scientists, Urey explained that the committee felt HUAC to be a disreputable and unconstitutional committee, and that it posed as great a threat to science as fascism or communism: "Attacks on scientists begin with scurrilous remarks but I believe that they end, both in Nazi Germany and in Communist Russia, by the murder of scientists—as well as other people."[59]

In September 1948, Urey was implicated in the HUAC investigation of the analytical chemist Clarence F. Hiskey. Hiskey had been chosen to work on the Manhattan Project based on Urey's recommendation, and HUAC accused Hiskey of being an active member of the Communist party and of giving atomic information to a Soviet espionage agent.[60] Based on Urey's associations and his outspoken political views, Howard Rushmore, a reporter for the New York *Journal-American*, argued before the Illinois Seditious Activities Investigation Commission that Urey should be barred from continuing to work on atomic projects, and that "an educator who could not 'discern' the true character of a Communist-front organization should be prohibited by law from teaching."[61] Urey responded to these allegations with the argument that secrecy and witch hunts only hurt American atomic research and, as a result, contributed to Soviet success.[62]

In 1948, the Emergency Committee even considered challenging HUAC head-on by putting Harrison Brown up for Congress in the Second District of Chicago against incumbent Richard B. Vail, a Republican and leading member of HUAC. The committee believed that "Brown's nomination would change the campaign from a local congressional election into a general fight of the atomic scientists of America against the methods practiced by the Un-American Activities Committee, for maintaining civilian control of atomic energy, and for a constructive foreign policy instead of Vail's isolationist demagoguery."[63] For a reason not documented in the archives, the committee chose not to pick this particular fight.

The attacks continued. As the physicist Sir James Chadwick summed it up in 1953: "Urey was very badly treated by the American authorities."[64] When in 1950 he was scheduled to give a speech in Helena, Montana, Montana Congressman Mike Mansfield contacted the FBI requesting information on "the loyalty of Dr. Urey" after "a number of individuals have questioned his Americanism."[65] The FBI had already investigated Urey and found him to belong to several Communist front organizations.[66]

The events of the Cold War reinforced Urey's view that religion was essential to the proper functioning of science. His activities with the Emergency Committee had often brought him before religious audiences, and Urey had seen how readily religious organizations joined in the struggle against the proliferation of atomic weapons. Moreover, he had become increasingly convinced that moral courage was what separated heroes from villains in the Cold War. This issue came to the fore in a 1949 letter to Rabbi Louis Finkelstein, in which he compared two university presidents, Chicago's Robert M. Hutchins and Harvard's James Bryant Conant. These two men were, to Urey's mind, polar opposites. Hutchins, a minister's son, had proved to be an unwavering champion of the scientists' movement. Conant had failed to defend scientists and educators from interference in the postwar years, and displayed a lack of intellectual courage. "[Conant] has consistently taken the line of least resistance in all of the problems facing education at present," Urey wrote, "including the very difficult situation involving loyalty oaths and investigations for subversive activities, while [Hutchins] has consistently maintained a very courageous stand."[67]

In his defense of Condon, Urey sometimes compared the complacency of scientists in the face of HUAC and McCarthyism to the failure of German scientists to stand up to the Nazis in their rise to power:

> Twenty years ago, Nazism began its rise in Germany, and, with some notable exceptions, German scientists did not stand up very well to this rise of tyranny. It came so insidiously. Unpopular people and ideas were attacked, and it was so convenient to look the other way and be busy with one's very, very, important tasks. . . . The state claimed that it was punishing criminals, but it became the chief criminal. Most of our scientific friends looked the other way.[68]

Urey explained to Finkelstein what he believed to be the root cause of such complacency. According to Urey, what Conant and these other scientists lacked was religion:

> We are living in a time when it is necessary that people stand up and be counted. The trend toward Fascist-Nazi tendencies in this country is really alarming. In Germany and Italy only the religious groups had the courage to stand up and be counted, and it is also true today that only the religious groups exhibit this courage insofar as activities behind the iron curtain are concerned. I am exceedingly disappointed in scientists generally, but I could not possibly be in favor of giving awards now to people without this kind of courage, and I think Conant does not have it.[69]

Urey was perhaps overly critical of Conant here, and this most likely reflected his own personal dislike for Conant (who was, after all, one of Urey's major antagonists during the Manhattan Project). Urey may even have resented Conant for his political savvy, a quality Urey mostly lacked.[70] There is some truth to Urey's complaints, however. Conant's handling of individual attacks on intellectual freedom was inconsistent, and he did attempt to influence the president of the National Academy of Sciences, Alfred N. Richards, not to issue a public statement condemning the attacks on Condon. As his biographer James Hershberg concedes, Conant had an "overly abstract, timid or aloof approach to defending academic freedoms"—he preferred to push for proper, legal investigations rather than to argue that investigations were unnecessary, unwarranted, or politically motivated.[71]

Urey's concern with the failure of scientists to stand up for one another was genuine. He did not understand why other eminent scientists like himself did not speak out more often. If he was willing to suffer investigations and public criticism to stand against the persecution of others, why were others not? Why did some seem to care more for their own political survival than for the good of science and democracy? Urey was now convinced that the loss of religious character in the scientist or the intellectual could have grave consequences.

A TROUBLED START

For the first decade after the end of the war, Urey attempted to balance his politics and aversion to weapons work with his scientific research and its reliance on government and military funding. (How he struck this balance on the scientific side, and how he built his postwar research program, is detailed in the next chapter.) On the political side, he had moved from advocating world governance to supporting weapons stockpiling in the name of peace. In his scientific work, he had found ways to build on his atomic expertise without working directly on weapons-related research. It was perhaps a fine line, but Urey walked it effectively, most of the time. However, this did not mean that he remained free from controversy or investigation. Urey's confrontations with HUAC and the FBI were exacerbated when he became involved in the highly publicized espionage trial of Julius and Ethel Rosenberg, who were convicted of conspiring to pass atomic secrets from David Greenglass, an Oak Ridge and Los Alamos employee, to the KGB. The case was brought to Urey's attention by a woman who went to see him speak against a congressional candidate at a Jewish temple in Chicago. The woman later appeared in Urey's office at the university and left some literature about the case with him. "Approximately two days later a transcript of the [court] record appeared on my desk," he later wrote to Joel Hildebrand.

> I am a curious person so I read it. It took a week of evenings to go through this. Half-way through I thought they were guilty as h——. I finished reading the transcript. I was shocked at the type of proceeding that passed for justice in a law court of the United States. I was shocked by what passed for evidence. I was shocked by the complete lack of any pretense of judicial objectivity toward the accused. I was frightened that such proceedings could take place in the United States. You will remember this was at the time of the McCarthy hysteria and the Korean War.[72]

Urey had little success convincing his colleagues in Chicago of the weakness of the state's case against the Rosenbergs. Still, Urey was convinced that the Rosenbergs' conviction and imminent execution were miscarriages of justice that would come to embarrass the United States. In December 1952, Urey sent a letter to President Harry S. Truman asking that their death sentences be commuted, writing, "I believe the Rosenbergs are or have been Communists or very sympathetic to Communist ideas. I regard such people as unreliable generally, but I do not believe in punishing people unless they commit crimes."[73]

Urey did not allege that the Rosenbergs were innocent, however. As he wrote to Condon only one month after his letter to Truman, he was "most unhappy to be mixed up in the Rosenberg case, and they may be just as guilty as hell, too."[74] Rather, Urey felt that the trial itself had been a violation of due process, that the prosecution was based on perjured testimony, and that the death sentences were too severe. On June 12, 1953, one week before the Rosenbergs were to be executed, Urey sent a telegram to President Dwight D. Eisenhower. Urey had failed to secure an appointment with the attorney general, and now pleaded for a personal audience with the president himself, during which he could present his understanding of the case, the improbability that the Rosenbergs could have passed atomic information to the Russians, and his skepticism as to the reliability of David Greenglass's courtroom testimony.[75]

The FBI and HUAC took a renewed interest in Urey after his support of the Rosenbergs. In 1953 the FBI interviewed Urey's Hyde Park neighbors, only to find that he "apparently leads a very quiet scholarly life."[76] His colleague and friend, the physical chemist Willard Libby, told the FBI that Urey was not disloyal but simply "naïve and innocent concerning political matters and that this innocence combined with Urey's renown as a scientist has caused many groups to seek him out either as a speaker or a supporter for their activities."[77] While Libby defended Urey's championing of justice in the Rosenberg case, and was sure that Urey would "never knowingly reveal any classified information which was entrusted him," he told the FBI that he "would conclude that Urey is so naïve and innocent as to constitute a security risk" if he joined a group to attempt to release the Rosenbergs' co-conspirator, Morton Sobell.[78] When HUAC published its report on the Rosenberg and Sobell affair, they characterized Urey and his many unwitting ties to Communist organizations as "a significant illustration of how the various Communist organizations, each created to accomplish a specific purpose, form separate but unified parts of the Communist conspiratorial system. . . . No clearer example [than Urey's] could be desired of how the long-range plans of the Kremlin, and the individuals recruited to implement them, maintain their continuity despite a maze of forms and names."[79]

CRISIS OF CONSCIENCE

The execution of the Rosenbergs in June 1953 was followed only three months later by the death of Urey's moral compass, his mother, Cora. Urey's Royal Society biographers do not mention her death, but they do state that it was in this year that Urey's hopeful attitude toward the improvement of society changed: "The execution of the Rosenbergs left him with a severe mental trauma. He did not consider the atomic secrets that he, himself, had helped to create sufficiently important to justify such a barbaric punishment."[80] Surely the Rosenbergs' fate was just the last straw, as Urey's optimism for world improvement seems to have been waning since the failure of the Baruch Plan.

It was in this year that Urey introduced a new speech to his repertoire that he titled "The Intellectual Revolution." This speech—versions of which he presented for the next ten years—contained a more developed view of the history of science and its effects on society than he had ever presented before. It was a story of two worlds: that of the scientist and that of the public. While the scientist produced knowledge at an increasingly rapid pace, the public only slowly absorbed this new knowledge. This growth in scientific knowledge had produced a drastically different view of the universe and humanity's place within it, and when this knowledge was fully absorbed might change "our philosophy of life and our ideas as to what men and women are" to a degree that matched that of the Reformation. It was a subject that needed to be addressed, Urey felt, because the failure to adjust properly to these changes had led countries such as Germany and Russia to adopt pseudoscientific societies based on flawed understandings of science.[81] This language echoed the language of the Conference on Science, Philosophy, and Religion's critique of totalitarian pseudoscience from a decade earlier.

Urey's intellectual revolution began with the introduction of the heliocentric theory of the solar system proposed by Copernicus, refined by Kepler, and systematized by Newton. The revolution continued with a summary of the great developments of the nineteenth century. First was the rise of chemistry "from almost one of the black arts to a great and exact science," with the introduction of the elements and the exploration of their properties, and culminating with the emergence of the periodic system of the elements. Next came the discoveries in electricity and magnetism, the properties of light, and the laws of thermodynamics. Most revolutionary of all, Urey said, was the discovery of biological evolution. All these new developments had helped reorder the world of human beings, and some had even caused science and religion to come into conflict. "[The] 'conflict of science and religion,'" Urey wrote, "can be more nearly dated from the publication of 'The Origin of Species' in 1859 than from any other event."[82] However, Urey also claimed that "the authors of these ideas and developments were themselves mostly very sincere and devout followers of organized religion and in no way intended to disturb, much less destroy, the religious beliefs of their time."[83] These conflicts were only "storms in teakettles or at least not more than summer thunderstorms."[84]

The great discoveries and developments of the twentieth century,

Urey believed, deserved much greater concern. First among these was the discovery of relativity and the development of quantum mechanics. Unlike the discoveries of the nineteenth century, which Urey believed were obvious truths previously obscured by superstition and hubris, these more modern discoveries were "the result of the most careful and penetrating analysis into the ultimate structure of the universe in the regions of its smallest and largest manifestations."[85] The same was true of the development of the science of heredity and the discovery of DNA (reported by James Watson and Francis Crick only two months before Urey presented his speech; Urey had nominated this work for the Nobel Prize). Perhaps most dramatic was how the discovery of radioactivity allowed geology to determine the age of Earth, confirming that it was billions of years old and that human beings had existed on Earth for only a relatively short time. In addition, astronomy with modern telescopes had by that time shown scientists that the Milky Way galaxy was vast—fifty thousand light-years across—and that the universe was infinitely vaster. Copernicus had ousted humans from the center of the universe, but these modern developments had shown humans that they were likely just one among millions of conscious and intelligent life forms that populated a universe where planets such as Earth might exist in great numbers.[86]

It bothered Urey, he told his audiences, that "these things have been accepted by the general public as the amusing speculations of scientists and as having little import for the ordinary human being."[87] Meanwhile, another society existed within which the developments of twentieth-century science were treated as matters of fact around which were constructed new philosophies—that of the scientists themselves.

Just as in his Depression-era speeches, Urey presented scientists as a group that practiced a special way of life that demanded a rigid code of ethics. Urey explained, "Scientists must be honest people with respect to their scientific work and it is this rigid honesty that is responsible for the great advances of science. No misrepresentation of facts or dishonest interpretation of them is tolerated and men who engage in such practices become ostracized by respectable scientists."[88] But what accounted for this upright behavior? Before the war, Urey had attributed it to the self-sacrifice and sublimation that objectivity required of the scientist. After 1953, however, Urey's ideas changed.

In 1953, Urey turned to religion. Urey resurrected the argument of the importance of religion for democracy that he had made a decade earlier among his allies in the Conference on Science, Philosophy, and Religion. He applied it now to science and survival in the atomic age. "Scientific training is not responsible for this honesty of approach to scientific problems," Urey said, "for a person's scientific training is largely secured long after the foundations of character have been established. Scientific training only selects the objective, honest person."[89] Instead, the source of the rigid code of ethics that scientists exemplified was the community within which science was practiced. The ethical code that underlay the community in the West was that of Christianity. Urey explained,

> It is . . . interesting to note that modern science has originated and has flourished in that part of the world where Christianity has been the dominant religion. . . . This religion and other religions emphasize the greatness of men, their very great capacities to understand and assume responsibility rather than their mere animal characteristics, and the greatness of the universe, and they admonish men to think of great things and to act in great and noble ways. The Ten Commandments of the Jewish religion have been brought to our world by Christianity, and it is this religion which civilized the Roman world and the savages of ancient northern Europe who are the ancestors of so many of us. "Thou shalt not lie." You must not lie about or misrepresent your data. "Thou shalt not steal." You do not assume that you have done work which others have done. These two commandments are of paramount importance to science and I do not believe that science can originate nor be maintained in a community which does not generally subscribe to and practice them. It seems to me that science developed in Europe because of the important influence of Christianity on the people of that continent.[90]

Like the Brethren thinkers of his childhood, Urey now made the argument that the teachings of the Bible were responsible for human progress, and drew from an understanding of community and practice not far from the one with which he had grown up.

Urey was skeptical that science would continue to flourish in the Soviet Union, where Christianity had been abandoned and replaced by "apologies to dialectical materialism."[91] He did not expect to see the quality of science in Russia decline within his lifetime, though: "It required centuries for Christianity to civilize Europe and it will require many years for the false doctrines of communism to destroy the remnants of the western tradition within that country."[92] However, his rhetoric in the years to come made it clear that he believed science could succeed only in societies where scientists were raised to act "as a good Christian should behave."[93] If Christian morals were eroded or eliminated from society, the decline of science was inevitable.

Scientists, unfortunately, could not help carry on the Christian tradition. Although they behaved as good Christians, most of them, Urey lamented, were skeptics who worshipped the universe. Urey could act as a good Christian and could observe the importance of Christianity, but he could not give nonscientists a reason to believe. Moreover, he worried that the general public took the wrong message from the success of science and engineering. They cared more for the practical applications of science than for the grand view of the universe it provided, and were mistakenly becoming amoral materialists. The solution to this dilemma could not come from science.

Science could provide an honest and even inspiring view of the universe, but "it gives little to the ordinary nonscientific citizen which enables him to meet the spiritual needs of life."[94] Only religion could meet these needs. The problem was potentially severe, given the uncertainties of the Cold War world, within which it was "so necessary that some inner well of strength be stirred and maintained for all men as individuals, for most occasionally, for some continuously."[95] Urey ended the first version of this speech with a plea to his audience that appealed to the two foundational texts of his childhood religion: "The drift from that high and moral life taught to us in the past must be arrested and we must not think that this scientific and engineering century can be built strong and true in any respect without adherence to the virtues taught in the Ten Commandments and the Sermon on the Mount. I urge all of you to try to fit the new concepts of science into the ancient teachings of religion to which we owe so much."[96]

CHAPTER SIX

A Return to Science

Even if his enthusiasm for isotope separation had waned, and even as he suffered the attacks of HUAC, in these immediate postwar years Urey was still optimistic about the future of nuclear science. Urey took on an unofficial leadership role as one of the Institute for Nuclear Studies' most senior and eminent members. Along with Willard Libby and Joe Mayer, Urey held a good deal of responsibility for the working atmosphere of the institute. In 1946 these three Berkeley alums initiated a Thursday afternoon seminar at the institute that closely imitated the colloquia they had been required to attend as graduate students under Gilbert N. Lewis. If Lewis had failed to invite Urey to join his coterie, Urey by now had established his own. Here Joe Mayer, who took the lead in the seminar, played the role of Lewis and picked from the day's attendees who should present their research. According to Hutchison, "No speaker or program was announced beforehand. The discussion was entirely spontaneous and informal. The blackboard, not slides and transparencies, was used. Sometimes there were two or three at the blackboard commenting on each other's equations or graphs."[1] Any kind of science that seemed interesting was discussed at these seminars, and the discussion was never discipline-specific. The researchers were pushed to apply their atomic expertise to the widest range of problems they could imagine. Often it was Urey who would end up dominating the discussion—either presenting his ideas about whatever articles he had read most recently, or critiquing whatever ideas others had brought up that day. In this way, Urey's postwar aimlessness was allowed to express itself in a productive setting.

At the end of 1946, still in search of a new line of active scientific work, Urey prepared and delivered that year's Liversidge Lecture before the Chemical Society of the Royal Institution, London. The Liversidge Lecture was one of Urey's last remaining prewar commitments. In this lecture Urey chose to update the isotope exchange equilibriums that he and L. J. Greiff had calculated and published in the 1930s, this time using a more sophisticated method developed for the SAM Lab by Jacob Bigeleisen and Maria Goeppert Mayer. Urey and Greiff had shown that relatively large differences in the physical and chemical properties of isotopic compounds could be detected—differences that were then exploited in the various separation techniques he developed in the intervening years. Revisiting the thermodynamic properties of isotopes now, with his postwar aversion to separation, Urey's mind latched onto another way that these chemical differences could be exploited.

In one section of his paper, Urey discussed the geological abundances of the isotopes of carbon and oxygen. Here, Urey noted that certain processes in nature tend to result in isotope enrichment. Aquatic carbonate-precipitating organisms, which use oxygen in their metabolic processes, tend to concentrate oxygen-18 (the more common of oxygen's two heavy isotopes) preferentially. The shells of these organisms often contain up to 4 percent more of the isotope than their surrounding waters. This enrichment is temperature sensitive, Urey's tables suggested, with a change in 25°C resulting in a change in the O-16/18 ratio of 1.004 relative to the water. "These calculations suggest investigations of particular interest to geology," Urey commented.[2] He further speculated that, with a precise mass spectrometer, a researcher could determine the O-18 ratio of carbonate rock samples to within a small degree of error, and possibly discover the temperature at which the rock was deposited with a certainty of within 6°C or less. Although Urey admitted that there was still a great deal of experimental investigation left to perform before the method could be put to use, he felt confident that oxygen isotope abundances were well suited to determining historic temperature changes. He concluded his paper by stating that the same small differences in the thermodynamic properties of isotopes and their compounds that "make possible the concentration and separation of the isotopes of some of the elements [in the laboratory] . . . may have important applications as a means of determining the temperatures at which geological formations were laid down."[3]

Although Stephen G. Brush's account of the postwar rise of geo- and cosmochemistry speaks to Urey's ability to attract researchers to the fields that he pioneered, in the beginning of his work on paleotemperature Urey seems to have had difficulty finding younger scientists to work in his new program.[4] The first postdoctoral fellow Urey did attract was Samuel Epstein, a young Canadian with mass spectrometer experience. As Epstein later remembered it, even though Urey had already publicized his speculation about the possibility of using isotopes in carbonate rocks to determine paleotemperature, there was no one lining up to work with him on the problem. When Epstein arrived, he became part of a small research team of four, consisting of himself, Urey, an engineer named Charles R. McKinney, and a graduate student named John McCrea. Once research got underway, Epstein found a different Urey than that described by Hutchins, Mayer, and Suess: "It was a joy to see Harold make a comeback in the scientific academic world. He never walked up a set of stairs one step at a time, always two steps at a time. His enthusiasm for his research was contagious. I clearly remember him coming into the laboratory dressed meticulously in a white shirt and coming home with a shirt stained with oil because he couldn't resist the temptation of changing a dirty oil pump or some other work that was usually left to the younger set."[5] Now feeling at home in the institute and excited again by what promised to be a fruitful research program, Urey was able to leave behind the traumas of war work.

BUILDING MASS SPECTROMETERS

In Ronald Doel's history of the American planetary astronomy community, he emphasizes the significance of mass spectrometers in the birth of isotope geochemistry as a source of the prestige and authority that Urey and his institute colleagues brought to the geosciences.[6] Mass spectrometers—which vaporized, ionized, and propelled chemical samples through an electromagnetic field, separating and then detecting the ions based on their mass—were the most sophisticated and precise instruments available for the measurement of isotope ratios. Not only did these instruments become more widely available after the war and accessible to scientists who had never used them before (biologists and geologists among them), but the type of work that Urey and Harrison Brown initiated at the University of Chicago would not have been possible without these instruments. However, while it is true that the Chicago isotope geochemists were among the instrument's first users after the war, it would be a mistake to assume that they had automatic access because of their wartime work.

Urey knew just how far mass spectrometers had developed during the war. While Urey may not have contributed much to Dunning's work on the gaseous diffusion method, he had successfully made his case to the army and the Office of Scientific Research and Development that it would require the manufacture of reliable mass spectrometers. The K-25 facility at Oak Ridge, which came online in spring 1943, pushed uranium hexafluoride gas through a series of porous barriers and separated the lighter isotope from the heavier one based on the relative ease with which the lighter isotope traversed the barriers. The gas was corrosive, and even a small leak could lead to big problems, as this would gum up the barriers and force the entire system to shut down. Urey insisted to Vannevar Bush that special mass spectrometers should be used as leak detectors, and he enlisted his colleague Alfred O. Nier from the University of Minnesota to adapt his existing design of a sixty-degree sector mass spectrometer for this purpose. Before the war, this type of instrument existed only in a handful of labs, built by the few experts like Nier who knew the necessary physics and engineering to build them. By the war's end, dozens of these leak detectors were mass-produced by General Electric for the Oak Ridge plant.[7] One hundred more mass spectrometers, also of Nier's design, were built by GE for uranium isotope analysis and were used at Oak Ridge to monitor the gaseous diffusion and electromagnetic process streams.[8] The placement of these machines within the Manhattan Project helped refine the mass spectrometer in both design and use.

Nothing was automatic about the transfer of this technology from its wartime service to its postwar diffusion into academic laboratories—even at Chicago's Institute for Nuclear Studies. Although Urey and his collaborators did believe before the war's end that the transfer of this technology would occur smoothly and without any great lapse in research, bringing the new generation of mass spectrometers online in American laboratories took at least two years of concerted effort. Making these instruments a postwar reality required Urey to wield much of the clout he had built up as a Nobel laureate and a Manhattan Project alumnus, and to sell the instrument's usefulness to agencies that had not traditionally funded research in atomic science.

Urey's claim in his Liversidge Lecture that oxygen isotope ratios could help unlock geologic temperature records was predicated on the availability of very sensitive isotope-ratio mass spectrometers. The instruments would have to be able to detect the small differences in oxygen-18 concentration that would result from temperature changes. Urey was confident that the mass spectrometers designed by Nier during the Manhattan Project had the necessary precision. Even before his research priorities had shifted from separation to the development of new geochemical methods, Urey was already imagining that he would have access to Nier's machines at the institute. When he sent his estimate of the costs involved in creating an isotope separation program to Bartky, Urey had expressed his belief that some of the setup money could be saved by getting Nier's instruments directly from the army.[9] This belief was reiterated in a letter to a biochemist colleague, in which Urey speculated that the mass spectrometers used during the war might be given second lives in university labs, where among other things they could assist analytically in biological research involving isotope tracers.[10] But despite Urey's optimism, the instruments were not quite so easily or cheaply obtained. Even as he put his research team together and approached funding agencies for the money he needed to operate his new program, he began to worry that this lack of equipment might keep him from getting his program off the ground.

Urey's expectation that the army would be willing to sell its mass spectrometers, or that they would at least be willing to hand one over to a private firm so that the instruments could be reproduced for sale to researchers, turned out to be unrealistic. Nonetheless, he and Nier were determined to make the instruments available in one form or another. In September 1946, after the wartime design of the spectrometers had been declassified, Urey wrote to A. V. Peterson, an engineer for the Manhattan District Research Division, about the prospect of having the army send a mass spectrometer on temporary loan to a private firm. Although Urey was not primarily concerned with tracer research, he pointed out to Peterson that these instruments would be useful in such peacetime research, and that they would soon be sought after for this purpose: "I myself would like to get an instrument of this kind, and I have no doubt many others would like them also."[11] Nier wrote to Peterson to reinforce Urey's request, and also promised "that in the event that the Manhattan District could temporarily release one of the instruments, I would be very glad to cooperate with anyone who could be found to manufacture other instruments of the same kind."[12] After the Manhattan District and the Patent Office of the OSRD studied the proposal, the two scientists were told that the army had no intention of letting go of their instruments, and knew of no existing procedure that would allow them to do so.[13]

Urey by now had anticipated this negative response. In an earlier letter to Nier, he expressed his doubt that going through the army would yield any result, and encouraged Nier to think about other routes that might bring the instrument to market. He lamented that the lack of reliable analytical instruments was "the greatest impediment to the use of stable isotopes," the very field that he was eager to promote.[14] Urey and Nier did find a way to use their prominent positions in the world of isotope separation to encourage the instrument's postwar development. They aligned themselves with a National Research Council committee on cancer research that wanted to see stable isotope tracers used in medicine: the Committee on Growth. In July 1946 Nier applied to the committee for an annual grant in the amount of $17,000 to be used in the construction and operation of a thermal diffusion plant for the production of carbon-13 at Minnesota. The grant would be used to supplement already existing and planned facilities and would, Nier promised, allow him to provide the committee with a portion of the carbon-13 produced in the plant that was proportional to the committee's support.[15] As the head of the Committee on Growth's Committee to Negotiate Purchase of Stable Isotopes, Urey approved the grant and encouraged the committee to pay Nier as quickly as possible so that he could begin work immediately.[16]

In a December 1946 report to the committee, Nier stated that the funds received from the grant were being used for two primary purposes: the separation of carbon isotopes by thermal diffusion (as stated in the original application), and the development of a new mass spectrometer to be used in tracer research. Nier also stated that the instrument was already in operation, that it was able to perform isotopic analysis of carbon, nitrogen, oxygen, and hydrogen, and that with some modification it could also be used for other elements or for general gas analysis. And he reported promising commercial prospects: "The work has progressed so far that I have attempted to find a company which will make a complete instrument so that it may be generally available."[17] Nier had in fact already identified a company, the Consolidated Engineering Corporation (CEC) of Pasadena, California, and had begun collaborating with CEC's Harold Washburn. In January 1947, two men from CEC visited Nier's lab for several weeks, and by February indicated that they would begin production of the instruments as soon as possible.[18]

Nier also wanted the instrument to be available to labs that could not afford a commercially produced instrument, and for this purpose prepared a paper for the *Review of Scientific Instruments*. He also offered the services of his equipment engineer to construct mass spectrometer tubes for a limited number of interested scientists, as well as detailed drawings, parts lists, and circuit diagrams. He worried that "many of the laboratories who could use an instrument of this sort will not be in a position to purchase a perfected commercial instrument," but hoped that "if given sufficient data could with some help put together an instrument which might not be quite as reliable or accurate but which would be adequate for many purposes."[19] All of this was with the intention of stimulating the growth of isotope research—a goal on which both he and Urey agreed. Reiterating his request for an extension of funding, Nier told the committee, "If this grant were received I would continue general development work along the same general line attempting to find ways to simplify the instrument in order that it could be built more cheaply and used more widely."[20]

As it turned out, Urey was one researcher who ended up building his own spectrometers rather than buy the CEC's manufactured spectrometers. Throughout the process of promoting the instrument's development, Urey had grown impatient. "I should like to get some mass spectrometers going here as soon as possible," he wrote to Nier. "I feel rather desperate after the war because I find that my colleagues and apparatus both have all been completely disorganized by the war activities. Sometimes I have thought that I never again will get started doing scientific work. I shall be very unhappy if that is the case."[21]

When a mass spectrometer was not immediately forthcoming, Urey decided to import mass spectrometry know-how into his laboratory, as he had in the 1930s. McKinney had worked as an electrical engineer for GE and spent the war in Oak Ridge maintaining Nier's mass spectrometers. Urey hired him to build an instrument that would work for his purposes. He also went looking for young researchers to work with him on his oxygen thermometer. After having no luck drumming up interest among possible young collaborators in the States, he contacted his former research assistant, the Canadian Harry Thode, to inquire if he knew of any talented young researchers. (Thode had assisted Urey in the mid-1930s with the design and operation of separation systems for isotopes of nitrogen, carbon, and sulfur. After working in Urey's lab, Thode returned to Canada, where he became a professor at McMaster University and later president of that university.)

Thode put Urey in touch with Epstein, then a 27-year-old isotope prodigy. Epstein sent along his credentials to Urey in summer 1947 and reported that in Thode's lab he was working on perfecting a mass spectrometer tube in which solids could be analyzed without having to be converted into gaseous chemical compounds.[22] Epstein was a competent physical scientist with a focus on the technical aspects of mass spectrometer research. This fit perfectly with Urey's new research goals. Less than one week later, Urey responded to Epstein's inquiry with an offer of $4,400 per year in salary, along with an apartment above the garage of his Hyde Park home for $60 a month. Urey also informed him that he would be working with McKinney, and that there would be two or three technicians to prepare the samples and perform the routine analyses in the spectrometer.[23] By February 1949, Urey's lab had constructed two mass spectrometers for the oxygen work.[24] The designs were based on the specifications Nier provided, as well as Thode's published descriptions of a sector mass spectrometer for isotope abundance measurements, modified for Urey's purposes.

NEW NETWORKS

Moving into geological territory meant that Urey had to develop a new network of scientific contacts and collaborators. In addition to Epstein, McKinney, and McCrea, he also called upon colleagues from Chicago's Department of Geology. Before the war, Chicago's geologists already had tended to be more lab-oriented than field-oriented. The Chicago geologists considered their geophysical program to be one of the strongest in the country, housing one of the only working high-temperature petrology labs outside of the Geophysical Laboratory at the Carnegie Insti-

FIGURE 10 Harold C. Urey at a mass spectrometer in his University of Chicago laboratory. Copyright 2019 The Chicago Maroon. All rights reserved. Reprinted with permission.

tution of Washington. The postwar period brought an influx of cash to the department. The university increased the department's typical expense and equipment budget of $1,500 per year to $45,000 for the first three postwar years.[25] This enabled the department to invest in new equipment and allowed for the conversion of some existing facilities into state-of-the-art laboratories for analytical chemistry.

FIGURE 11 Harold C. Urey examining marine shells and fossils, 1953. Urey used oxygen isotopes found in belemnite and other fossils to determine the temperature record of the oceans with his oxygen thermometer. Courtesy of the Special Collections Research Center, University of Chicago Library.

Starting in 1946, the Department of Geology, now feeling itself to be in competition with its counterparts in physics and chemistry, adopted the attitude that "anyone on the staff who was not opening up brand-new fields was a piece of dead wood."[26] To facilitate change, the department hired new faculty members, including the geochemists Julian Goldsmith, Hans Ramberg, and Kalervo Rankama. These men, particularly Goldsmith, would work closely with Urey, Libby, and Brown to bridge the gap between the Institute for Nuclear Studies and the Chicago geologists, and all would assist in proposing a joint curriculum in geochemistry for students who wished to become geochemists.[27] As early as 1947 the department was receiving contracts from the Office of Naval Research to do geophysical research. The navy even put some "paperclip specialists" (German scientists who had worked under the Nazi regime) under the care and supervision of Chicago's geology faculty.[28]

In 1947, Urey secured the cooperation of the German-born paleoecologist Heinz Lowenstam, who had left Germany before the war and was working for the Illinois State Geological Survey. In 1948 the Department of Geology hired Lowenstam specifically to work with Urey on his paleotemperature studies, and Lowenstam's salary was paid through Urey's research contracts.[29] Beginning in 1948, Urey asked scientists at the Scripps Institution of Oceanography in La Jolla, and at other marine laboratories, for shells and information about the waters in which they had been deposited. Epstein and Lowenstam worked to develop methods for preparing uncontaminated samples of carbon dioxide gas from calcium carbonate shells, then worked to establish a temperature scale for oxygen isotopes. The first published results of this work appeared in 1951.[30]

In 1950, Cesare Emiliani completed his doctorate in the Department of Geology and went to work with Urey's paleotemperature group. He extracted foraminifera shells from long deep-sea cores. Using those shells, the group studied temperature variations in the Pleistocene and estimated the length and severity of the ice ages. The acquisition of the deep-sea cores was evidence of Urey's diverse and expanding scientific network, especially his connection to the emerging earth science network. In 1950, Urey's lab began collaborating with Columbia University's newly established Lamont Geological Observatory, a "quintessential Cold War institution" that Columbia had established in order to take advantage of military support for geophysics research.[31] There, with substantial support from the Office of Naval Research, Maurice Ewing had developed a method for piston-coring seafloor sediment. Throughout the 1950s, Ewing and his colleague David Ericson sent core samples to Urey's lab, where Emiliani and the lab's technician, Toshiko Mayeda, prepared and analyzed the samples in the mass spectrometer. But, as an examination of Urey's research funding makes clear, Urey's connection to Cold War military contract research went far beyond his connection to the Lamont Geological Observatory.

FUNDING THE NEW PROGRAM

In its early years, Urey's new research program benefited from the close alliance between the institute and industry. As Ronald Doel points out, the petroleum industry was a major supporter of geophysical and geochemical research during the Cold War.[32] Both Shell and Standard Oil had bought memberships in the Institute for Nuclear Studies, for which they were promised the right of first refusal on any patents or practical applications developed there.[33] During a tour of the institute in 1947, a Shell representative met with Urey and heard about his research plans. The man from Shell came away from the meeting impressed. Writing to the chairman of the American Petroleum Institute (API) Advisory Committee on Fundamental Research on Occurrence and Recovery of Petroleum, he characterized Urey's research as "of considerable interest, since, if successful, it will help measure one more of the many unknown variables of importance to the origin of oil."[34]

Urey had given the API the impression that his research might contribute to the understanding of the processes that produced oil. In his initial courting of API funding, Urey offered this speculation: "It may be that oil deposits occur in places where the temperature at which they were deposited was unique in some way, and if this should be the case then it might furnish one additional tool for geological exploration for oil."[35] It is also possible that the oil companies were interested in developments in mass spectrometry generally, as the method had been introduced within the petroleum industry in the early 1940s and had proved highly useful as an accurate way of analyzing hydrocarbon mixtures.[36] Urey now requested $12,000 for the construction and maintenance of his instruments, but the API was willing to grant him only $5,000 for

1948/49.[37] That amount fell well short of what Urey estimated it would cost just to build his first mass spectrometers, much less do anything with them.

In summer 1947, Urey requested funding from the Geological Society of America's Penrose Bequest, playing up the possibility that his work would replace existing qualitative methods of determining paleotemperatures—namely, paleoecological studies of the fossil organisms found within geological samples—with more quantitative methods.[38] The Geological Society granted Urey $17,900 for salaries to support one chemist, one physicist, and three technicians.[39] That amount, even when combined with the API funding listed above, still didn't approach the $50,000 to $100,000 that Urey estimated he would need in order to build all the necessary instruments and establish the new methodology.

The petroleum industry and the Geological Society of America supported Urey's work in these early years, but they were either unable or unwilling to provide the amount of money Urey needed in order to launch his new research program in earnest. Eventually they withdrew their support entirely. As the API explained to Urey, there were "several other more desirable projects which are basically fundamental in nature, but [were] still closer to [their] immediate problems" than was Urey's.[40]

Military patrons, however, were both willing and able to make the investment. In 1949 the amount of funding Urey had at his disposal increased dramatically as he began a new contract with the Office of Naval Research (ONR). Urey had participated as a scientific observer in the navy's Operation Crossroads atomic bomb test at Bikini Atoll in 1946. There he had met Roger Revelle, future director of the ONR's Geophysics Branch. He no doubt also became acquainted with the navy's attitude that "almost all fields of oceanographic research had potential Navy applications."[41] In 1948 Urey put his aversion to military contracts aside and made his first contract proposal to the ONR, asking for about $105,000 for an "investigation of natural abundances of stable isotopes with the primary objective of measuring paleo-temperatures."[42] The proposal was vague about the practical applications of paleoclimate research to the navy's mission, but Urey did manage to frame the more general aspects of isotope abundance measurements as having the potential to contribute to the navy's existing mapping program and to develop natural tracer techniques that could be employed in the ocean. The navy agreed to provide roughly $30,000 per year for four years.[43]

The Office of Naval Research was, in some ways, an ideal funding agency for the early years of Urey's research program. The navy preferentially funded the development of new methods and techniques. From their point of view, Urey's work might help them better understand the ocean's basic geochemical features and assist them in the development and maintenance of naval technologies. However, in 1952, once Urey's methods had been established, the ONR informed him that they were no longer willing to fund his research.[44]

The withdrawal of ONR funding put pressure on Urey to find a new funding agency to take its place. He was able to find two that together raised his funding to still greater heights. In 1953/54, Urey received $55,956 from the US Atomic Energy Commission and $21,400 from the National Science Foundation (which had established its Earth Science Program in 1953).[45] With more than $75,000 in contract funding, 1953/54 was a banner year for Urey's research program. In his remaining years at Chicago, before he left for the University of California at San Diego in 1958, Urey would keep his external funding at or slightly above this new level.

Although Urey had decided to leave isotope separation work behind, much of his clout with the Atomic Energy Commission and the military was attributable to his expertise in that field and to his past position as the head of Columbia's SAM Lab. For this reason, it was not only difficult but also impolitic for Urey to completely close the door on it. In fact, Urey had been involved in forming the AEC and had been working under contract with it since November 1950, first as a consultant on a "Heavy Water Production Processes Survey" for the AEC's Division of Research.[46] Remaining connected to the AEC's concerns about heavy water and isotope separation—and flexing his expertise in this area at its behest—helped Urey maintain the prestige he had earned from his wartime service. It also allowed him to keep abreast of the commission's concerns (even at times to define these concerns) and made it easier for him to frame his new projects in language that would garner the AEC's approval. This relationship was symbiotic: Urey received support for non-separation-related research; meanwhile, not only was the AEC satisfied that his new interests were close enough to theirs to merit funding, but it was able to enlist him in the work of advising and planning the AEC's activities. Unclassified projects such as Urey's also gave the AEC examples of supported research that could be discussed and promoted before Congress and the public.[47]

One example of this symbiosis at work is Urey's reluctant agreement to chair the Committee on Isotope Separation for the AEC's Division of Research in early 1951. In a letter to Kenneth Pitzer, its director, Urey wrote: "Long ago I developed a subconscious reaction to all separation jobs. It is, first, that any separation project is an enormous amount of hard and uninteresting work, and second, that it is very likely that all new schemes for separating isotopes will not work."[48] However, accepting the position allowed Urey to exert some influence on the direction of isotope work in the United States and put him in constant contact with Pitzer. Furthermore, the Committee on Isotope Separation had a high priority within the AEC's Division of Research. The committee was charged with recommending to the division what steps should be taken for the investigation and development of isotope separation techniques.[49]

In addition to keeping Urey and his fellow members of the committee connected to the Division of Research, the work kept them connected to classified materials and places of atomic research. As a contractor and a consultant, Urey maintained the security clearance that had been granted to him during the Manhattan Project. The AEC installed facilities in the offices of committee members for the storage of classified documents (if they didn't already have such facilities) and initiated clearance procedures for secretaries and technical assistants. The members also received a classified bibliography of sources held in classified libraries at the National Laboratories and the associated universities.

With his knowledge of the inner workings of the commission, Urey was able to construct proposals for isotope geochemical work that were directly related to the AEC's concerns. In 1949, Harrison Brown floated a "Proposed Program for the Accumulation of Quantitative Data Concerning: The Chemical Composition of Meteorites and the Earth's Crust; the Relative Abundances of Elements in the Solar System; the Ages of the Elements and Planets," and hoped that the AEC would at least fund those parts of the program that were performed at its Argonne facility. The AEC demurred. Urey's first proposal was far more politically savvy in both name and form. Urey's proposal for "Research on the Natural Abundance of Deuterium and Other Isotopes in Nature" outlined an intentionally broad research program that included work to be done on meteorites, igneous rocks, and fossils, with the stated aim of discovering how the abundance of hydrogen isotopes had changed over time. In addition to addressing the AEC's concerns about deuterium and heavy water and their abundances in nature, Urey's proposal also emphasized the scientific attention that his initial work on paleotemperature was receiving, thus tapping into the AEC's desire for visible scientific rewards from unclassified and nonmilitary projects.[50]

THE ORIGIN OF LIFE IN THE COLD WAR

Urey's Cold War concerns and his scientific research program were not always kept separate. One case of his origin of life experiments with his graduate student, Stanley Miller, illustrates this well. By the early 1950s, Urey's interest in geochemistry and new research in meteorites led him to consider the chemistry of the solar system and the origin of Earth. From this, Urey's interests wandered into the question of the origin of life. An article in *Science* sparked this interest.[51] It proposed that the chemistry of life had been formed in an oxidizing atmosphere by high-energy discharges, and researchers had used Berkeley's sixty-inch cyclotron to subject carbon dioxide gas to accelerated helium particles.[52] Urey felt that his former Berkeley colleagues' approach was "quite irrelevant to the problem of the origin of life."[53]

Near the end of 1951, Urey began preparing a paper of his own on the subject, and presented his preliminary argument before one of the institute's weekly seminars. Here he presented his point of view: that the early Earth had a highly reducing atmosphere consisting primarily of hydrogen, ammonia, methane, and water. Electrical discharges within this atmosphere, such as lightning, formed the first carbon compounds, establishing the prebiotic conditions from which life could emerge.

Sometime between Urey's initial presentation and the publication of his paper, the Russian scientist Alexander I. Oparin's hypothesis that life emerged under reducing conditions also came to Urey's attention. Urey incorporated Oparin's hypothesis as further support for his own claims. It is likely, however, that Urey was already set on a reducing atmosphere by this time. As he reported to the AEC, he had been led to this conclusion by his own team's finding that the igneous rocks of Earth had a far lower abundance of deuterium than average, and that meteorites had less deuterium than the mean of the oceans. "We interpret this latter result to mean that 10 or 15% of the present amounts of water on the earth has been dissociated and the hydrogen has escaped," Urey wrote to his patron, "leaving the deuterium mostly behind. This leads to the conclusion that the original atmosphere of the earth was highly reducing and that an oxidized atmosphere has been produced only by the escape of hydrogen."[54]

In his paper, Urey suggested an experiment on the production of organic compounds from water and methane in the presence of electrical discharges similar to lightning strikes.[55] He must have described a similar experiment in his seminar presentation: "I went back to my office, after the lecture, and Stanley Miller appeared. He said he wanted to do his doctoral dissertation on the subject."[56] Urey was at first hesitant to allow Miller to take on the project: "I was afraid it might not turn out well for a dissertation, but Stanley could not be discouraged." With Urey's guidance, Miller designed an experiment that would simulate the conditions Urey proposed. According to Urey, the experiment paid off quite quickly: "In three months he had made the experiments and had secured evidence of compounds characteristic of living things."[57]

In early 1953, Urey and Miller were ready to publish Miller's results. Urey now felt he faced a dilemma. Only one journal in the United States could bring such a groundbreaking and interdisciplinary experiment to the attention of all the relevant disciplines, and this was *Science*. But sending the paper to *Science*, Urey complained to Edward Condon, would mean having to wait months for the paper to be published. If he sent it to a more specialized journal, it would be published faster but would not have the impact Urey desired: "We wish to send off a brief note for quick publication about this. It is of interest to zoologists, chemists, biochemists, geologists, and astronomers. Where should we publish this in the United States? If I send it to Science it will be six months before it will get in print. If I send it to the [*Journal of the American Chemical Society*] it will be out in a few weeks, but will miss a large fraction of the audience."[58]

Urey did send Miller's paper to *Science* in February 1953. By March, *Science* had decided to run the paper as one of its lead articles. However, before *Science* could send Miller an acceptance letter, Urey became concerned that the editors were taking their time with an article that he believed contained one of the most dramatic discoveries of the century. Urey sent a telegram to *Science* on March 10, requesting that the magazine return the paper at once, and stating that it had been sent with the understanding that it would be given speedy publication. He ended the telegram angrily: "It is being sent elsewhere cancel my subscription to Science as I regard it as a useless publication."[59]

Urey's overzealous championing of his student's work almost prevented Miller's paper from being a lead article in *Science*. As the editor Howard Meyerhoff wrote to Miller, he had just been dictating a letter to Urey and Miller about the acceptance of their paper when Urey's angry telegram arrived. Luckily for Miller, Urey was not the lead author on the soon-to-be-famous paper. As Meyerhoff explained, he was happy to cancel Urey's subscription to *Science*, but since Miller was the lead author, he would return the paper only if Miller so desired.[60] In May, Miller's paper appeared in *Science*. Urey was not listed as an author, but a footnote thanked him for his helpful suggestions and guidance.[61] As Urey explained, "Well, after this, I did very little and let Stanley Miller develop the project."[62]

Allowing Miller to claim the origin of life experiment as his own achieved multiple goals. Most obviously, it established Miller as a leading voice in origin of life research. Less obviously, it allowed Urey to shelter Miller from his own Cold War problems. The experiment took place in the midst of the Korean War; Miller was, as a graduate student, not necessarily guaranteed draft deferment.[63] In the early 1950s, the army was experimenting with draft deferment for graduate students.[64] Miller, however, was classified 1-A, available for unrestricted military service. Only three months after Miller's paper appeared in *Science*, Urey wrote to Miller's hometown draft board to plead Miller's case. Miller, he told them, was a brilliant student whose experimental work on the origin of life had received a great deal of notice from scientists in many different fields. He also told them that if Miller's work was interrupted before he graduated in the next year, it would not only be difficult for him to finish his dissertation, but other scientists would carry on his work without him: "This means a very definite disadvantage to this young man unless he is able to finish his work, and I believe it is a disadvantage to the country as a whole to interrupt the advanced training of a capable man."[65]

This was not the only way that the Cold War intruded on Miller and Urey's work. Origin of life research was an international affair, and Russian scientists such as Oparin were in fact leaders in this new field.[66] In 1956 Oparin invited both Miller and Urey to come to a Symposium on the

Origin of Life in Moscow in summer 1957. At first Urey accepted the invitation and advised Miller to do the same. He wrote to Miller, "My advice after thinking it over for some time and talking to several people, is to accept the invitation. . . . One never knows what this will do to us sometime in the future but I think it is safe."[67] However, soon the violent Soviet reaction to the Hungarian revolution changed Urey's mind. Miller wrote to Urey that the subjugation of the Hungarians had shown the Soviets to be "gangsters."[68] He speculated that it might not be career suicide to attend the conference with the endorsement of the State Department, but "even with endorsement, there is the question of whether I want to have anything to do with people like the Russians. It is often said that science should be separate from politics, but in this situation I am not sure."[69] Urey, whose wounds from the Rosenberg trial must still have been fresh, agreed with Miller's assessment: "I am very put out by the behavior of the Russians and do not feel like doing anything to indicate any acquiescence in their treatment of Hungary; but to the Russians I probably will not say this since I think it would do no good."[70]

CHAPTER SEVEN

To Hell with the Moon!

With his students and colleagues in Chicago, Urey had studied the isotopes of the light elements and introduced entirely new methods into the earth sciences. Once these methods were developed, and once Urey had trained a considerable number of students in their use, he decided that it was time to leave the field to his younger colleagues. Urey began searching for a research project of his own. The project he settled on would extend the methods developed in his Chicago lab to the cosmos, and would take on the daunting task of describing in chemical terms the formation and evolution of the solar system.

A few events conspired to push Urey in this direction. First was the 1949 publication of Ralph Baldwin's *Face of the Moon* by the University of Chicago Press. This was a book that Urey might not have read had he not been a member of the university's Committee on Publications in the Physical Sciences. Urey, who was new to the committee, was sent a review copy of Baldwin's book after publication. Dutifully, he took the book with him on a trip from Chicago to a speaking engagement in Canada, and by his account spent a good ten hours carefully reading the book on the train and trying to understand every detail completely. As he told one correspondent, the trip was to a Canadian nuclear power plant, and the book gave him some respite from the nuclear issues that plagued him: "On the way, I read Ralph Baldwin's book on the moon and became immensely interested in the moon as a result of that. Then I forgot all about the Canadian power plants."[1] Indeed, although Urey would continue to engage in nuclear debates, in 1950 he "quit" the Atomic Energy Commission, claiming that the work bored him, and told reporters that he was "more interested in my present work than in anything I was able to do in connection with the development of the bomb."[2]

Baldwin's book summarized the history of lunar geography, described the different types of lunar features, and evaluated the various hypotheses put forward to explain these features. He developed a strong argument in favor of the meteoritic-impact hypothesis, and presented his hypothesis for the history of the Moon and its maria. Baldwin had been trained as a physicist, but he had left his career as a scientist to work as an executive for the Oliver Machinery Corporation in Grand Rapids, Michigan. Urey was not impressed with Baldwin as a scientist. He found him to be unfamiliar with standard scientific notation, and thought that his math skills left much to be desired. As Urey said in an interview, "Now I'm not an awfully good mathematician myself. There are just many people that are much better than I at this. . . . But [Baldwin] is far worse than I am."[3] Urey was able to forgive Baldwin any perceived shortcomings, though, because he felt that Baldwin treated the competing theories and hypotheses respectfully and did not neglect to mention theories or theorists with whom he disagreed. After returning to Chicago, Urey asked the press for Baldwin's address and struck up a correspondence almost immediately.[4] The two men continued to correspond and share ideas and frustrations about lunar issues until 1976.

Baldwin's book compelled Urey to take a closer look at the Moon. He requested pictures from the Harvard astronomer Harlow Shapley, pasting these images together in his Chicago office to make a composite map of the Moon that was "about a meter across and was a pretty good picture of the moon."[5] It was from this composite that Urey formed his initial idea of it as a cold and ancient object: "What impressed me . . . was almost certainly [that] the principal features of the moon were due to great collisions. . . . The face of the moon predominantly was a picture of the final stages of the formation of the Earth—a history that was not available from the rocks of the Earth."[6]

Still, although his interest was piqued, and while he would continue to maintain that the Moon held special relevance in the history of the solar system, it was not the first planetary body Urey tackled. In the same year that Baldwin's book was published, Urey was asked to coteach a Summer Session course with Harrison Brown, "Chemistry in Nature." Urey's first lecture for the course was on the heat balance of Earth. In order to prepare for this lecture, Urey turned to a review article by Louis B. Slichter. Here he learned for the first time that the temperature of Earth had been rising, not falling, throughout its history.[7] Urey's encounter with Slichter's paper led him within a matter of months to new considerations of the role of radioactivity in the movements of Earth's crust and mantle. In the autumn of that year, Urey presented his theory of the cold accretion of the terrestrial planets, within which relatively homogeneous masses of material from the solar nebula were slowly heated by radioactive uranium, thorium, and potassium. This radioactive heating melted the iron within each planetary mass, which further heated the planet as the heavy metals flowed inward to form the core and displaced the lighter materials that formed the mantle and crust.[8]

Urey's move into planetary science continued in late 1949 when he attended the Conference Concerning the Evolution of the Earth in Rancho Santa Fe, California, cosponsored by the National Academy of Sciences and the University of California's Institute of Geophysics. Slichter arranged the conference in reaction to Urey's paper on the origin of Earth's core.[9] Not many of the participants in the conference were willing to accept Urey's model in its entirety. Many were, however, eager to accept Urey's method of initiating the gravitational separation of heavy and light elements through radioactive heating, thus allowing for convection in the mantle. Urey's model solved what had become a crisis in the field—explaining how Earth's iron core had formed in the first place, and how mantle convection was initiated. Earth scientists already accepted an iron core, because it explained seismic and magnetic data. However, wartime work on the behavior of solids and fluids under high temperatures and pressures, along with postwar inquiries into the viscosity of inner Earth, indicated that gravitational separation of light and heavy elements would take several billion years. That Urey was able to jump-start this system through radioactive heating, followed by gravitational heating, seemed to solve at least this one pressing problem.[10]

Even if Urey's planetary hypothesis won few adherents, he was able to test some of his early ideas about the formation of Earth and other terrestrial planets at this conference. As this was one of the first interdisciplinary planetary conferences—involving physical chemists, geochemists, geophysicists, geologists, physicists, and astronomers—it also allowed Urey the opportunity to incorporate ideas from these disparate fields into his own thinking.[11]

Another event that pushed Urey down the planetary path was the departure of Harrison Brown from Chicago. In 1951, Brown left the Institute for Nuclear Studies and moved his research program to Caltech, taking Sam Epstein and Heinz Lowenstam with him. Urey had been interested in starting meteorite research earlier but had felt it was inappropriate to intrude into his younger colleague's field. Once Brown decided to leave, however, Urey would no longer be stepping on anyone's toes if he picked up meteorite studies in his own lab. He allowed Epstein, Lowenstam, and Cesare Emiliani to take the paleoclimate work as their own and ended his work in this field. At this same time, the Austrian physical chemist Hans Suess, then with the US Geological Survey, spent the academic year of 1950/51 as a fellow at the institute and a guest in Urey's lab. Suess had only recently (in 1949) begun writing scientific papers in English, but had nonetheless already established a firm reputation in nuclear geochemistry.[12] Together, Urey and Suess began working on the abundances of the elements.[13]

Urey also struck up a collaboration with Gerard P. Kuiper, astronomer at the University of Chicago's Yerkes Observatory and president of the International Astronomical Union's section on planetary research. Urey incorporated Kuiper's model of the early solar system, within which a massive solar nebula formed large protoplanets through gravitational collapse of the nebular material.[14] By spring 1951, when he was to deliver the Silliman Memorial Lectures at Yale University, Urey felt that he had tested his ideas sufficiently and gathered enough evidence to contribute something substantial on the subject of cosmogony: the science of solar system formation. These lectures, published as *The Planets* in 1952, amounted to a chemical treatise on the origin and development of the solar system.

In Kuiper's model, planetary systems were formed in special cases of failed binary star formation. In the place of a companion star, the stellar nebulae in these instances produced protoplanets. Kuiper's model had chemical problems, and after the Rancho Santa Fe conference, he incorporated Urey's ideas to solve them. Kuiper adopted the chemist's hypothesis of the cold accretion and subsequent radioactive and gravitational heating of the terrestrial planets, and in turn Urey adopted the astronomer's binary star and protoplanet model. But this relationship went sour quickly. In the early 1950s Kuiper began moving away from Urey's cold accretion model and toward a hot model of planetary formation in which the terrestrial planets had gone through a completely molten phase.[15]

That Urey had to depart from Kuiper's theory in order to keep his model of the Moon intact became clear to him as early as 1955, when Kuiper published a paper in which he claimed that the Moon had at one point been completely melted. It also became clear to Urey that year that the planetary community was split between the cold and hot models.[16] In addition to publishing criticisms of Kuiper's model, Urey began writing to colleagues in astronomy and geology to try to convince them of Kuiper's errors. For example, he wrote to the astronomer Dinsmore Alter, "I do not believe [Kuiper's] is a tenable position at all and I think that all his conclusions are colored by his theory, so I doubt his observations very much."[17]

In a critical review of Kuiper's work, Urey asserted that although the astronomer had based his arguments on new telescopic observations of the lunar surface, in fact he had not observed "anything markedly different from what [had] been previously observed and recorded in the extensive literature on the moon."[18] As in his letter to Alter, Urey claimed that the evidence Kuiper now saw was actually derived from Kuiper's theory. Urey felt that his own theory, on the other hand, was based on unbiased observations. The most compelling of these was the Moon's apparent lack of isostatic equilibrium—the detection of gravitational anomalies indicated that the mass on and beneath the lunar surface was not evenly distributed. He interpreted this as evidence that "the moon cannot now or ever have been in the past as plastic as the outer parts of the Earth are today. This means a cold origin for the moon, and it means that it has remained cold up to the present time."[19] Urey also was convinced that the age of the solar system as determined by Clair C. Patterson, together with his own determinations of the abundances of the radioactive isotopes, presented "serious difficulties" that the hot model could not overcome.[20]

The split with Kuiper was not limited to differences of theory. Urey was also at this time feuding publicly with the astronomer, whom he accused of using his ideas without citing them properly and attempting to pass them off as his own. As noted earlier, in 1953 he had begun arguing in his public speeches that the Ten Commandments laid out clear norms for the practice of science: "'Thou shalt not lie.' You must not lie about or misrepresent your data. 'Thou shalt not steal.' You do not assume that you have done work which others have done."[21] He told his audiences that any scientist who violated these commandments would be driven out of the scientific community. Now, only two years later, feeling that Kuiper had transgressed these norms, Urey began a letter-writing campaign to have the astronomer removed from the presidency of the International Astronomical Union's section on planetary research and to have the National Science Foundation subsidies for his editorial projects canceled.[22]

One might conclude that Urey was growing cranky in his old age. However, at least some of the blame for this feud did fall on Kuiper. Even his protégés and collaborators agreed that he was a difficult man to get along with. "He was very, for want of a better word, authoritarian," wrote George Coyne. "He was of the European school of 'the professor says, and the students do what the professor says' kind of thing." Alan Binder recalled that "very few people could tolerate being with Kuiper. . . . There's less than a dozen people who can say, 'I studied under Kuiper.' To me, it's all a matter of Darwinian evolution: Either you could stand being under him, or you couldn't. Very few people really could."[23] His biographer for the National Academy of Sciences, Dale Cruikshank, put it: "As an intensely driven man, Kuiper's perceived hauteur occasionally strained the patience and loyalty of his colleagues and friends."[24] Personality conflicts with his colleagues would eventually lead to a state of "civil war" at Chicago's Yerkes Observatory and ultimately to Kuiper's departure.

Urey began forming his own model of solar system development in 1956. He decided to work backward from the physical properties and chemical composition of meteorites to reconstruct what kinds of processes and objects had formed them. From this exercise, Urey concluded that two types of objects had been created and destroyed before the accretion of the planets. These objects, which he called "primary" and "secondary" objects, were of lunar and asteroidal size, respectively. From the ages of the meteorites, Urey determined that the primary objects must have accumulated 4.5 billion years in the past. In order to melt the metal and silicates within the primary objects, they had to be heated by some means to the melting point of these materials, and then cooled to 500°C for several million years. After cooling, the primary objects were broken into very small fragments (ranging from a millimeter to a centimeter) and then reaccumulated to form the secondary objects. The breakup of these secondary objects then formed meteorites.[25] The primary and secondary objects thus predated Kuiper's protoplanets, and it was the pieces of these original objects that accreted to form the planets.

Urey continued developing his model in 1957, while in England as the Eastman Professor at Oxford University. Here he completely abandoned the binary star model and made the primary processes in the formation of the planets chemical rather than astrophysical. His model of solar system formation made planets a normal result of the birth of stars. Unlike Kuiper's, his model did not require special astronomical circumstances. Instead, Urey postulated that, as any great rotating mass of gas contracted to form a new star, the need to preserve angular momentum caused it to throw off a cloud of dust and gas. Most of this cloud would flatten into a disk rotating in a single plane about the newly formed star. Near the star, a denser portion of the cloud would shade the rest of the disk, absorbing much of the radiant energy of the star and allowing the disk to cool. The primary objects formed from this disk and then underwent their subsequent transformations.[26]

Meanwhile, Urey made moves to distance himself physically from Kuiper. In August 1957, Urey wrote a letter to Willard Libby detailing the reasons he no longer wished to remain in Chicago, and asked his colleague if he knew of "a possible 'out.'" First among the reasons that he listed was that, ever since Kuiper had been appointed the director of Yerkes Observatory, it meant for Urey "no satisfactory contact with the astronomers from now on. This has been the case from 1952 and was somewhat the case since 1949. . . . If I had been a younger man, I would have left the university and my present field of research long ago."[27] Libby wrote to the oceanographer Roger Revelle, who was then recruiting for his new university in La Jolla, California, and explained Urey's unhappiness. Within a matter of months, Urey had accepted an appointment as professor at large of chemistry at the University of California.[28] By November 1958, without having yet found a buyer for their Chicago home, Harold and Frieda Urey had moved to La Jolla to join Revelle's newly formed University of California, San Diego.[29]

THE GREAT PROPHET AND THE COSMIC NARRATIVE

Urey's turn from atomic work to cosmogony was not lost on the press, who were enamored by the idea that the chemist had taken up the cosmic epic as his new subject. As early as 1949, after he published his cold accretion hypothesis, the *New York Times* applauded Urey, who "knows all about radioactivity and makes proper allowance for it," for taking on cosmogony—work they labeled "scientific romancing."[30] Scientists, the *Times* wrote, "are supposed to concern themselves only with facts"; the history of cosmogony, however, from Kant and Laplace to T. C. Chamberlin and F. R. Moulton, proved that "when it comes to wild romance they eclipse the most extravagant fancies of those who contribute to 'pulps' given over to science fiction."[31] The *Times* understood Urey's compulsion to provide an accurate and scientific description of what had historically been the realm of religious folklore and philosophy. About Urey's "dream which is concerned primarily with the earth," the *Times* wrote:

> Here we have a sample of the folklore of a scientific age. A primitive savage could explain the wind only by supposing it was a blast from the mighty lungs of an invisible demon. The sun and the moon were similarly personified. Today we tell the same tale with improvements. We have the old stage, meaning the heavens, but the characters of the play, the stars, wear different costumes and talk a different language. Electrons, protons and neutrons strut about where once there were spirits. Instead of Greek gods on Olympus we have Greek symbols in equations. The wonder of how it began, the dreaming, is still there. And why not? Creation—there is no theme so stupendous. Only a bloodless dullard would fail to speculate about it. Let's have more fiction of the type that Dr. Urey has given us. There is something epic about it.[32]

This notion that science was rewriting the folklore of humankind and providing a new view of the universe against which to understand the human drama certainly fit with Urey's emphasis on the universe itself as a potential source of quasi-religious inspiration.[33]

But inspiration was one thing; meaning was quite another. The problem of retaining the traditional morals of Western religions had remained central to Urey's public speeches. As the *New Scientist* reported in 1957, Urey was a man of science, but "from time to time he lays down his papers on the origin of the world and indulges in oratory, that he may not have to witness the end of the world."[34] In the late 1950s, Urey added a new element to his earlier plea to keep science and technology in check with religious values, and argued that "one of the pressing needs of this age is a great prophet who can accept the facts of science and at the same time can give inspiration to fill this great void."[35] In the colleges of Oxford, where he spent the 1956/57 academic year developing his new model of solar system evolution, Urey reported that he had seen "the enormous change in our religious attitudes" to a greater extent than he had seen anywhere else. "How does religion maintain the old values? Can it do so without the miraculous, and without illogical dogmas? Can it make use of the magnificent view of the universe supplied by science and the materialistic necessities and luxury supplied by its applications to give us a sound moral life and noble aspirations?"[36] He assured his audiences that the religious men of Oxford were striving daily to find the solutions to these problems.

Urey's attitude toward God during this time indicates that he was no believer—at least, not in any traditional sense. Urey did not speak or write often about his own view of God. When asked whether his scientific education had eroded or strengthened his faith in God, Urey might respond as he did to one admirer's letter: "I myself have my own definitions of God and things of this sort, but I would not like to discuss them in public at all."[37] In his autobiography he admitted that he did not believe in an afterlife, and believed instead that "we are all temporary, and are only part of this enormously complex universe that changes with time."[38] In a public forum, after reciting from memory the first chapter of Genesis, Urey said that it was "beautiful poetry," but that he was more convinced by the evidence for evolution.[39]

When Urey did invoke God, it was a God who had done little more than set the universe in motion. When asked in an interview whether his disbelief in a personal God meant that he believed the creation of the universe was an accidental event, Urey answered, "No, not at all. . . . I follow the astronomers and their hypothesis of a Big Bang 50 billion years ago."[40] Urey preferred to believe in a "God [who] opened his hand and the universe was created."[41] As he explained to the editor of *This Christian Century*, it was simply "more beautiful to have a universe that is established in such a way that it takes care of itself completely by itself than it is to assume otherwise."[42] And as he told one audience, the universe was like a god: "Now this God is a God that extends in all directions for billions of light years and has existed for billions of years and will exist for billions of years in the future and maybe all these numbers should be infinite. This God has left a true record in nature which we can read with exercising some diligence."[43]

It was the universe that inspired Urey: "To me, the enormous universe and all the things in it are the source of my wonder, and I need no God to increase this wonder at all."[44] Urey was happy believing that the universe, beginning with the Big Bang, had proceeded to unfold and evolve without intervention, and that life had emerged on Earth through chance chemical events. But he also wondered whether something might not be at work underneath the seeming randomness of the universe. Urey pondered the question of "whether anything in nature is chance at all. It seems to me that the properties of organic compounds are such as may have resulted in the spontaneous evolution of life."[45]

So, for the most part, Urey was an agnostic. He defined himself as an atheist because he did not believe in "a private God that listens to the prayers of anyone."[46] However, he also believed that nature was too complex to be fully understood by the human mind. "Now, what do we do with this enormous heart of nature which is beyond any possibility of our comprehension?" he asked. "It looks to me as though this is an unknown and uninvestigatable region and it will always be impossible to decide whether the things which we observe in nature are naturally so or whether they are guided by an outside intelligence."[47]

As much as Urey believed that he and his fellow scientists didn't need a god to feel the wonder of the universe, he also believed that scientists alone could not provide a religion for the atomic age—and that humankind would not survive the atomic age without a religion. However, with the help of a "new prophet"—with religious thinkers who were willing to substitute science for superstition—they could still contribute. Because this new religion would require the most complete and inspiring view of the universe possible, scientists in fact had an essential role to play. Not to participate in this vital task of giving life meaning could have dire consequences even for the most skeptical intellectuals, Urey said: "It is necessary that the old fashioned morals be maintained for the daily well being of all men. It is necessary for science that these be maintained if it is to remain vigorous and active and if it is to continue to widen the horizons of men. It is necessary for intellectual pursuits of all kinds that the ancient moral teachings be maintained in an age of science."[48]

A FAILED PROPHET

Ruth Nanda Anshen—philosopher, author, and self-described "intellectual instigator"—took note of Urey's work on the Moon and his growing concern over the loss of religion.[49] If anyone could have taken on the role of the prophet that Urey described in the 1950s, Anshen was a fine contender. And if anyone may have reinforced for Urey the notion that his planetary work might be valuable to the religious thinkers of his time, it was Anshen. She had been a graduate student of the philosopher of science Alfred North Whitehead at Boston University in the late 1930s. Under Whitehead's guidance she had developed a preoccupation with what she decried as the "atomization of knowledge."[50] She retrospectively described her life's work as having been guided by a desire to "lead man from an age of fragmentation to a new plateau of consciousness."[51]

As a response to atomization, Anshen developed a "lifelong obsession with 'the unitary structure of all reality.'"[52] In order to illuminate this unitary structure, she set out in search of a "thematic hypothesis" or "unitary principle" under which all aspects of life and knowledge could be organized.[53] In the early 1940s, she initiated a series of edited volumes titled the Science of Culture Series, with an editorial board of Einstein, Bohr, and Whitehead. Within this series she collected and edited texts from some of the twentieth century's most influential scientists, philosophers, theologians, and authors. Einstein reportedly was very enthusiastic about the project, telling Anshen, "[It is] a very good plan. You want the future to come sooner."[54] By the late 1970s she had founded five additional series and through these had published more than 130 edited volumes and monographs. These series became "a crusade to ferret out kindred spirits, those European and American thinkers who shared her vision."[55] Many of those whom she ferreted out were former participants in the Conference on Science, Philosophy, and Religion, at which Urey had first expressed his view of the civilizing force of Christianity and its importance in preserving democracy. As Anshen herself described the purpose of these endeavors, they were similar to those of the conference (although not limited to the American way of life): "We were heretics, burning not at the stake but in our hearts and minds with one unending plea for unity. . . . We hoped to provide the core of a cultural Magna Charta for the guidance of our species."[56]

When she was first planning the Science of Culture Series, Anshen contacted Urey and asked if he would contribute to what would become her second edited volume, *Science and Man*. Eager to participate, Urey at first promised Anshen an essay that he said would be titled "Chemistry and the Physical Foundation of Civilization."[57] By January 1941, however, Urey asked "to be excused from doing so," explaining that he had been so busy with his war work that he had even had to cancel all his scientific lectures for the year.[58] Instead of writing a new essay, Urey contributed the transcript of a speech he had given in 1937, "The Position of Science in Modern Industry."[59]

After the war, Anshen shared Urey's concern that a "fact-based, materialistic world" would be "perilously devoid of ethics and values."[60] It was her contention that this very "threadbare" worldview was responsible for the destruction of Europe during the war. Not long after *The Planets* had appeared in print and Urey's public rhetoric had shifted in the direction of his comments at the conference, Anshen approached him a second time in the hope that he would be willing to publish a monograph in her new World Perspectives Series.

Anshen recognized a kindred soul in Urey. She told him that she was approaching a select few scientists who shared her viewpoints on the importance of integrating science with spirit:

> Only those spiritual and intellectual leaders who possess full consciousness of the pressing problems of our time are invited to participate in our Series: those who are aware of the truth that beyond the divisiveness among men there exists a primordial unitive power that we are all bound together by a common humanity more fundamental than any unity of doctrine; those who recognize the error of the environmentalist who forgets that man is an element of every experiment and that the most important element in man's environment is his fellowmen; those who realize that the centrifugal force which has scattered and atomized humanity must be replaced by an integrating process and structure giving meaning, purpose and dignity to existence.[61]

For his contribution, Anshen suggested that Urey publish an abridged version of *The Planets* that would be understandable to the general public. She tentatively titled this hypothetical volume *The Origin of the Solar System*.

Urey was at first enthusiastic about the possibility of publishing such a book. He instructed Yale University Press to send Anshen a copy of *The Planets* and agreed to a deadline of September 1955. He told Anshen that he was in the middle of attempting to revise the book for Yale and bring its discussion of the available data up to date, but that he felt this would not take long; he was confident that the general arguments he had made in the book were still valid.[62] Anshen was also enthusiastic to have the distinguished chemist onboard. After she had read the copy of Urey's book that Yale sent to her, she wrote to him that "perhaps the new consciousness of our epoch consists in the increasing awareness of the cosmic influences on man. How reassuring it is to know that you contribute to this consciousness with so much lucidity and integrity."[63]

Urey never met the 1955 deadline. Although he and Anshen continued to correspond for the next thirteen years, and even met in person twice to discuss the importance of bringing Urey's new work to the public, the book they envisioned never appeared. Over the next thirteen years, Anshen continued to ask Urey for a monograph about the planets. In language that sounds quite similar to Urey's own public rhetoric from this period, Anshen suggested, "Our series is dedicated to the definition of what may be considered a revolution in thought and a widening of horizons comparable perhaps to the beginning of the new scientific era. . . . This revolution seems to be taking place in spite of the intransigence of nationalisms and in spite of the spiritual and moral erosion of our time."[64] She assured him that his contribution would help "restore sapientia to scientia and to articulate the integrating forces which are moving in the two hemispheres of humanity."[65] She further implored him, "It is indeed indispensable to the spiritual and intellectual community that your seminal thought be represented."[66] Urey continued to tell Anshen that he needed more time to revise his original work and provided her with new estimated dates of completion, but nothing ever came of his promises.

UREY'S MOON IN THE SPACE AGE

> It is obvious that Congress is not appropriating five billion dollars a year for science. The object is adventure. People wish to put a man on the moon and bring him back. They talk about other things but they are not the true reasons for the program. Of course there is competition with the USSR, and as a matter of fact, this is very valuable. It even induces our own NASA to do some interesting scientific work from time to time. . . . My enthusiasm for the space program is just one of trying to see that we learn something during the course of this enormous project.
>
> HAROLD C. UREY, "AFFORDING THE SPACE PROGRAM"

Despite Urey's interest in the planets, he was neither an immediate nor enthusiastic supporter of the US space program. When the launch of Sputnik shocked American citizens and politicians out of complacency, Urey's response was not to call for an equally ambitious space program. Like other colleagues, he was not convinced of the inherent value or scientific merit of space exploration. Instead, Urey practically scolded the United States for its materialism and its general neglect of science, remarking, "Sputnik is a salutary lesson for us if we learn from it. We can afford defense, education and scientific achievement if we wish to do so. We can drop our waste of resources, and manpower, on gaudy and oversized cars, for example, if we wish to do so."[67] Clearly, Urey still pined for his days of bicycling through the streets of Copenhagen, or even further back to the days of cycling from Kendallville to Cedar Lake.

In order to counteract American waste and put resources to good use, Urey perhaps drew on his austere Brethren upbringing and proposed a 10 percent sales tax on all automobiles and luxuries in the next year, with 20 and 30 percent taxes in the years to come. Rather than putting this money directly into a space program, Urey believed, it would be better placed in new school buildings, increased teachers' salaries, and basic research. He insisted that his faith in democracy was unshaken, although he did not pass up the opportunity to criticize McCarthy and HUAC: "It was not communism . . . that [put the USSR ahead in the space race], except insofar as people worked, as apparently they have under the system. And it is not democracy which is at fault in the west, except insofar as it seems to have bred a contempt for the intellectuals, professors of our universities and our schoolteachers and instituted witch hunts against some of our most intelligent and patriotic people. Let's grow up!"[68] The Soviets had prevailed, he told his audience, because they "respect their educators and scientists and hold their intellectuals in high esteem. They support their schools and research establishments in a pre-eminent way according to reports which I believe to be reliable."[69] If Americans did not match the Russians in these areas, the Cold War was lost.

Urey's criticisms of the space race continued. When the National Aeronautics and Space Administration came into existence in fall 1958, a local Chicago newspaper asked for Urey's thoughts on the proposed space program. Urey responded with "a most negative interview." He "did not believe that the program was worthwhile and . . . had no interest in sending men to the moon." Deciding that he did not want to sound so negative of a program he felt would inevitably develop with or without him, Urey contacted the newspaper the next day and asked that they not run the story: "My reason for making this request was that I was sure that when men acquired the capacity to go to the moon they would go to the moon whether I thought it was worthwhile or not."[70] But Urey would become very invested in the space program, and especially in missions to the Moon, within a matter of months.

Urey's lack of enthusiasm may have stemmed at least partially from the fact that the majority of the scientists and administrators who made up the new NASA were either atmospheric scientists, military personnel, or engineers. These people had joined the American space effort early on because of their interest in sounding rocket research, national defense, and aerospace issues. According to R. Cargill Hall, "sky science" clearly held the upper hand in the new space program.[71] However, planetary scientists had gained some leverage within NASA by the end of 1958. Largely this was due to the efforts of Robert Jastrow. The new assistant director of Space Sciences, Homer Newell, brought Jastrow—a physicist from the Naval Research Laboratory's upper-air research group—to NASA in November 1958. When Newell tasked Jastrow with forming a Theoretical Division, Jastrow sought out the existing community of planetary scientists—many of whom were represented at the Rancho Santa Fe conference—and brought them inside the NASA fold. Newell described Jastrow as "a prime mover in regard to academic ties," an "imaginative theorist," and a "superb speaker."[72] Within NASA, Jastrow's ability to "hold both lay audiences and professional colleagues spellbound with his descriptions of space science topics" was a boon when it came time for a representative of the agency to appear before Congress and defend the annual budget request.[73] He also became one of the agency's greatest public advocates—appearing on television and publishing several popular articles and books about the ways in which NASA was benefiting humankind.

Urey's enthusiasm for constructing a cosmic narrative aligned with Jastrow's interests. The two also had in common a shared agnostic desire to examine the relationship between science and religion—which Urey was already articulating in his public speeches and Jastrow would later express in popular books like *God and the Astronomers*—although this does not seem to have been the basis for their working relationship.[74] According to Jastrow, one of the very first things he did upon joining NASA was to seek out Urey. Jastrow had decided that, instead of focusing on every scientific problem that could be studied by satellites and rockets, he should organize his division around "a few important problems."[75] Jastrow decided that Urey, since he "had written a book on the moon and the planets, and was well known for the intensity of his interest in the scientific study of these objects," would make a good ally within the scientific community.[76] That he was a "great American man of science" and had written not "a dry discussion of the solar system, as such books usually are," but one that "was enlivened by a sense of evolution in the Cosmos and the place of our planet in the larger scheme of things," made Urey even more appealing as one who could sell the program to the public.[77] By Jastrow's account, he immediately saw the potential benefits of bringing Urey into his division. When they met in person, Jastrow was impressed by Urey's ability to explain his theory of the Moon's origin:

> [Urey] sat me down, handed me his book on the planets, opened to the chapter on the moon, and began to tell me of the unique importance which this arid and lifeless body has for anyone who wishes to understand the origin of the earth and other planets. I was fascinated by his story, which had never been told to me before in fourteen years of study and research in physics. Harold Urey had the marvelous quality of an intense, almost childlike curiosity regarding all aspects of the natural world. This kind of curiosity is a rare quality.[78]

Jastrow claimed that he recognized in Urey "that cosmological spark" that could make space science interesting.[79]

In *The Planets*, Urey had in fact laid out one possible scenario within which the Moon might be of unique significance. This was a scenario in which the Moon formed independently of Earth during the early years of the solar system and was captured by Earth after accretion. This was a scenario that, Urey admitted in his book, was rarely given serious consideration because "[the moon's] orbit is not retrograde, not highly inclined to the plane of the ecliptic, and not very eccentric."[80] But he also noted that his own review of the chemical evidence did not rule out the capture scenario, and that if the Moon formed from its own protoplanet and then was captured by Earth, this could explain the large angular momentum present in the Earth-Moon system as well as the chemical differences between the two bodies.[81]

Due partly to his ongoing feud with Kuiper, as well as to meteorite work done in his Chicago lab, Urey's theory of the origin of the Moon and solar system had changed since the publication of *The Planets*. By the time Urey sat down with Jastrow for the first time in 1958, he had begun revising his theory of the origin of meteorites. The general story of the primary and secondary objects remained intact; however, the secondary objects were now the parents of only the chondritic meteorites. The stony meteorites, on the other hand, Urey now believed must have come from the Moon or from other Moon-like objects. This led Urey to the conclusion that the Moon was in fact a primary object, older than Earth, that had been captured by the planet after accretion. Urey speculated that there were likely far more remnant primary objects in the solar system in the past, and that they may have played an important role in the evolution of the solar system. As for the Moon, Urey argued that it should be "composed of the more nearly correct solar average material of the less volatile kind than the earth and other terrestrial planets."[82]

Urey explained his new theory that the Moon might hold the key to unlocking the secrets of cosmogony to Jastrow. As opposed to Earth, Urey believed that the interior of the Moon was cold, and that it had been cold for most of if not all its history. Whereas most of the rocks on Earth's surface are young, those on the Moon would be much older. Furthermore, if the Moon had been cold since accretion, it would be completely undifferentiated—meaning that the rocks on the surface of the Moon would be just as old as any rocks in the lunar interior. As evidence that the Moon had never been hot, Urey pointed out that the Moon had a frozen tidal bulge that would have collapsed had it ever had a soft interior.[83] Jastrow quickly latched on to Urey's conclusion that the surface of the Moon would hold the record of its birth, or at least of the early years of the Earth-Moon system. It gave Jastrow a compelling way to sell a role for space science in missions to the Moon: "It could tell us something we would never learn on the earth; it could help us solve the mysteries of the origin of the solar system and the origin of life."[84]

Jastrow resolved that he would use Urey's theory of the Moon's origin and significance in order to make the case for the inclusion of planetary research amid the already existing emphasis on Van Allen belts, orbiting telescopes, and human missions into Earth orbit. Jastrow invited Urey to come to NASA in January 1959 to give lectures on the Moon and planets to an audience of space scientists and NASA administrators. Afterward, Jastrow and Urey met privately in Jastrow's office, where Jastrow complained that "the Russians were wiping up the floor with us in space." According to Jastrow, it was Urey who then suggested that NASA step up its existing plan to land a craft on the Moon in 1963. He reportedly asked, "Why don't we get on it before then and show the world we can do something scientifically important in space."[85]

When Jastrow and Urey approached Newell about advocating a crash project to land on the Moon by 1961, he requested that the two men write a memo to NASA's director of the Office of Space Flight Programs, Abe Silverstein. Jastrow wrote the political part of the memo while Urey handled the scientific aspects. The memo argued that a crash lunar project with a real scientific agenda would "enhance the reputation of the United States to a degree that cannot be achieved by the execution of a conventional scientific program on a normal schedule," and that "a soft landing with performance of [a few basic experiments] will capture the imagination of the scientific community and the general public to a greater degree than any project of comparable scientific value."[86] The memo also stated emphatically the authors' belief that a lunar study was of greater scientific importance than a study of Earth's other near neighbors, Venus and Mars, which was based on their claim that "there is written plain to our eyes on the surface of the moon the history of the origin of the solar system."[87]

Lunar proposals were nothing new. Since even before the founding of NASA, proposals for missions to the Moon had been circulated among the leadership of America's then mostly military space effort. The earliest of these was a 1957 proposal by William H. Pickering, director of the Jet Propulsion Laboratory (JPL) in Pasadena, California. Pickering and Lee DuBridge, president of Caltech, agreed that lunar flights were the most appropriate response to the Soviet threat. With DuBridge's support, Pickering proposed "Project Red Socks," outlining a series of nine rocket flights to the Moon.[88] When the Advanced Research Projects Agency formed in February 1958, its director, Roy Johnson, agreed that lunar missions would be an appropriate response to Sputnik. He announced in March of that year that the agency would make the evaluation of American capability to explore space "in the vicinity of the moon, to obtain useful data concerning the moon, and to provide a close look at the moon" one of its primary activities.[89] The air force also began its Pioneer program in 1958, the early missions of which were meant to prove that a study of the Moon with rockets and probes was possible. The rationale for these programs was clear: the Moon represented a tangible goal, and one that the Russians had not yet reached. The scientific benefits of such a program, on the other hand, were not specified.

Urey's was the first statement drafted within NASA that put forward a scientific rationale for lunar exploration, and it was certainly the first to rank the Moon above Mars and Venus in scientific importance. Newell remembered that Jastrow and Urey used their memo to "[undertake] a small campaign to sell the idea to NASA."[90] For his part, Newell saw the potential of such a campaign and passed the memorandum up the chain of command to Silverstein with his endorsement. According to Jastrow, "Harold Urey was the trigger, I was the bullet, and Homer Newell fired the gun."[91] The memo reached the highest levels of NASA and eventually made its way to Congress. By Newell's account, "Urey's story [of the Moon] provided good ammunition for moving the proposal on up the line. The persuasiveness of the argument carried the day at each stage, within NASA, in the Administration, and finally in Congress, and in due course investigation of the Moon was formally and officially a part of the NASA space science program."[92] Throughout the program's early existence, Urey would continue to provide much of the scientific justification for lunar exploration.

PROMISED THE MOON

These developments in the latter half of the 1950s would seem to indicate that Urey had a new platform from which to promote his Moon. His revisions of *The Planets* were going well. In September 1959, Urey reported to his editors that he had completed a revision of the chapter on the Moon, was in the process of revising the chapter on the terrestrial planets, and had written an entirely new chapter on meteorites. He wanted to distribute the revised chapters on the Moon and meteorites immediately in mimeographed form to his colleagues at NASA. He felt that this was essential because, as he put it, "there have been a number of very bad reports put out by various people . . . who have not studied the subject at all carefully and merely report that so-and-so says this and so-and-so says that without any attempt to evaluate ideas. . . . [My ideas] may not be right but at least one person has thought about the subject for 10 years and has attempted to put down his ideas in systematic form."[93] At this point, Urey still felt that his model of the Moon would be the driving force of NASA's lunar program.

Even as Urey remained excited about his new relationship with the space program, however, he also felt that he had great difficulty within NASA's lunar working groups getting a fair hearing for his ideas. He had succeeded in creating his own model for the formation of the solar system and the origin of the Moon as a primary object. However, he had failed to do away with Kuiper's model, or to win over many converts within the planetary science community. While the community was still split over whether the Moon was hot or cold, whether it was captured by or was ripped from Earth, and whether its features were due to volcanism or bombardment, Urey's chemical model was not the go-to model even for the cold Moon contingent. Furthermore, much to Urey's dismay, Kuiper was making inroads within NASA, and it seemed to Urey that the astronomer's ideas were being treated more favorably than his own.

Kuiper's 1959 theory of the Moon's origin was not substantially different from what he had presented in 1955, and Urey was annoyed that Kuiper had done little to take his criticisms into account: "I criticized this theory in 1955 and his reply was quite unsatisfactory. Kuiper may not like my criticisms but he has an alternative, namely, to stop talking about it."[94] A few years later, Urey wrote to Jastrow to complain, "What a dreary

FIGURE 12 Harold C. Urey and the planets. The final phase of Urey's career was focused on the origin and evolution of the solar system. He was especially drawn to the puzzle of the Moon, and to the idea that it might hold the chemical evidence of the history of the solar system. Courtesy of the Mandeville Special Collections Department, University of California, San Diego.

business the space program is in many ways. . . . One must be around a certain astronomer much to one's discomfort, etc."[95]

The move to La Jolla had represented a clean break with the Chicago astronomers. Thus it is not surprising that Urey was unhappy with NASA's in-house staff, their dismissal of his theories, their unfamiliarity with his own publications on the subject, and their uncritical acceptance of Kuiper's work. In one particularly discouraging exchange, Urey reported, an unnamed scientist from JPL visited him in La Jolla, "[sat] at my desk and [told] me essentially that he does not believe anything I say and apparently does not even think my arguments worth considering."[96] This prompted Urey to write an angry and severely critical letter to Al Hibbs, head of the laboratory's lunar projects, stating that he would rather not have his time wasted in this way.[97] In a later letter to Hibbs, in which he apologized for his initial outburst, Urey admitted that the true nature of the problem was the uncritical acceptance of Kuiper's theories at JPL:

> It does seem to me that someone connected with the program might try to evaluate a theory that is ten years old. Of course, I am to blame also because I too accepted the theories rather blindly. My only excuse is I am not a trained astronomer and I assumed that the work being published was being critically reviewed by others in the field and that I could trust the results of such studies. . . . Of course, part of my difficulty is that I am very much annoyed at myself for being taken in, but I am even more annoyed at the fact that there is no critical review of this subject.[98]

These complaints to Hibbs continued. In November 1959, Urey wrote to Hibbs about the lunar symposia, stating, "The people that run them do not know good men from bad. Therefore, the program always has poor papers and good papers mixed without any discrimination whatever."[99]

NASA was finding Urey to be a difficult man to manage. The Moon that Urey had helped Jastrow sell to NASA and Congress was scientifically unique, and it was this uniqueness that made the Moon valuable in Urey's eyes. Urey was comfortable stating in the memo he coauthored with Jastrow that a lunar mission would be of greater scientific importance than a similar mission to Mars or Venus because of his belief that the Moon would hold evidence of the formation of the solar system that had long since been erased on the other terrestrial planets. A high-profile scientific mission to such an important object was what Urey imagined would "capture the imagination of the scientific community and the general public to a greater degree than any project of comparable scientific value."[100] His version of the Moon put the secrets of the solar system in close enough proximity that samples could be gathered, and put isotope geochemists in the position to unlock those secrets. Furthermore, that this science could be done in view of the world, in the midst of a highly publicized space race with the USSR, would allow these chemists to perform publicly the type of inspirational work that he had described in his public speeches. He and his colleagues would in essence perform a public communion with the laws of nature.

Any other moon would not fit the bill. This became clear, to Jastrow's dismay, at the end of 1959 in a press conference held by his Theoretical Division. One reporter asked Urey why he was so insistent that NASA should answer important questions about the Moon, and not just show the world that the United States could reach it: "I was under the impression that there was some objective in reaching the moon, other than simply learning about it."[101] Urey responded that, although he was personally "thrilled by a feat of exploration," such as when explorers had reached the South and North Poles or when the Russians had photographed the far side of the Moon, from a scientific point of view there was no justification for such exploration if not to determine the Moon's origin: "Our primary concern as scientists is to try and understand this universe. I am much more interested in the origin of the solar system . . . than I am in making a trip somewhere."[102] When it came to the Moon, Urey thought he knew exactly what NASA should find. If something different were found, it would hardly be worth looking in the first place. Jastrow, on the other hand, while acknowledging the place of existing theories as a starting point for exploration, painted lunar exploration in the colors of territorial exploration. The Moon might hold the answers to Urey's big questions, but it was also the next great frontier.[103]

Urey's frustrations with NASA were amplified as his feud with Kuiper resurfaced yet again. In early 1960 Urey became aware that the astronomer was not only gaining favor within NASA but also planning to move west to take a position at the University of Arizona, where Urey feared he would come to dominate the newly constructed Kitt Peak National Observatory. Urey wrote to Albert Whitford at the Lick Observatory to see if he could enlist his support in an effort against Kuiper. On the same day, Urey sent a similar letter to Kitt Peak's director, Aden B. Meinel, asking if he did not feel "a little concerned about the possibility that the instabilities of Yerkes Observatory might be transferred to Kitt Peak." He told Meinel, "I think it is no great secret that many, many people who have done work anywhere near that which interests this man have had considerable trouble. Possibly it is always the fault of the other fellow but I doubt this to some extent." Urey emphasized that his concern was not due to any personal feud that he had with Kuiper, but about securing a "stable regime" at the country's newest and most advanced national observatory.[104]

Urey's overture to Meinel fell on deaf ears. Meinel was one of Kuiper's colleagues at Yerkes before coming to Tucson and was already well aware of the conflicts in Chicago. Meinel had headed the study that selected Kitt Peak as the location of the new observatory. Once he became director of the observatory, Meinel helped Kuiper extricate himself from his problems in Chicago. According to the account of one of Kuiper's collaborators, these problems arose not from any problems with Kuiper's personality but from "undercurrents of discontent [that] were circulating among some of the nonlunar-oriented personnel," and from Kuiper's "generally strong-arm tactics" in promoting and favoring lunar work over more traditional astronomical work.[105]

Meinel brokered an agreement with the University of Arizona's coordinator of research that would allow Kuiper to establish a NASA-supported Lunar and Planetary Laboratory at the university.[106] By the time Urey wrote to Edwin F. Carpenter, director of Arizona's Steward Observatory, explaining that Kuiper was the reason he had left Chicago, it was too late.[107] Kuiper had already visited Tucson and laid the groundwork for his new lab. Although Carpenter was no great fan of Kuiper, Meinel had convinced him that the astronomer and his NASA ties were worth whatever trouble he might bring with him.[108]

Urey's crusade against Kuiper was entirely unsuccessful. By the end of summer 1960, Kuiper and his staff were in Tucson. Also that same summer, Kuiper was invited by NASA to serve as a consultant on their Planetary and Interplanetary Sciences Subcommittee.[109] The die was cast. Urey's complaints continued, but he no longer seemed to expect that anything would result from his criticisms. Late in 1962 he wrote to his friend the Dutch astronomer Jan Oort about a recent lecture Kuiper had given in Leiden on the structure of the Moon. Urey had heard about the lecture through a colleague in attendance and was not happy with the report he received: "[As] usual he neglected the work of other people. . . . In this case even the technique he uses to produce the pictures you will note was invented by myself several years before he undertook to use it."[110] He sent Oort a reprint of the 1956 article in which he had originally criticized Kuiper's theories and asked Oort, as a fellow "product of The Netherlands," to account for Kuiper's behavior: "Perhaps you understand why it is that he spends so much effort in attempting to give the impression that previously published work is indeed original with him at much later dates."[111] Urey concluded his letter with one last expression of his exasperation with Kuiper, asking Oort, "Why don't you take him back to Holland?"[112]

Urey's enthusiasm for his relationship with NASA waned in the early 1960s, and his work suffered. Just days after his initial letters to Whitford and Meinel, he was reporting to Yale that "during the last three months I have gotten nothing done on the book. In fact, I have been pulled by the space program and its various meetings in one way or another until I have gotten nothing at all done. I am now trying to curtail these activities and if need be I will resign all connections with it in order to finish the job."[113] Urey's experience at NASA was on the whole "discouraging," and eventually completely soured him toward the idea of revising and publishing an updated edition of *The Planets*.[114] As he reported to his editors in 1961, trying to convince his fellow lunar scientists of his theory's merits was futile, not because his ideas were unsound but because everyone had a theory of their own: "I find that I have worked for years on this problem of the origin of the solar system and I have advanced ideas which seem so reasonable to me, but other people do not believe them at all. In the last year a number of other people have undertaken to write similar things; I have no doubt but that they think they are very reasonable ideas, yet I do not believe their ideas at all. In fact, I have a feeling that no one working in the field believes what anyone else does."[115]

Even within Jastrow's theoretical group, where he felt he could get a sympathetic hearing, Urey didn't have to look far to find unsympathetic responses to his capture hypothesis. He wrote to Jastrow toward the end of 1962 to complain about the behavior of his junior colleagues, two of whom had "published a paper in *Icarus* on the moon in which they disagreed with everything I have said about the moon during the last 10 years but they thanked me profusely at the end. This makes me thoroughly angry. I do not like to be thanked for being their 'whipping boy.' . . . I know your intentions are good but I do not like being 'made over' and then have my ideas ignored."[116] By summer 1963, Urey seemed even more demoralized than ever. He wondered "why I bother to do anything for NASA," and complained that the agency was "the most unsatisfactory contact I ever had . . . in my life." At every committee meeting he attended, Urey felt that his fellow lunar scientists only wanted to "clobber" his theories: "It seems to me that people think that everything I do is wrong; in fact, everything that I say in one of these committee meetings is wrong immediately as soon as it is said without any doubt."[117]

Urey likened his dealings with his younger space science colleagues to his experience with Dunning during the Manhattan Project. In an interview, he compared these young scientists to the scientists he produced in Chicago: "You find some young men [who] come along and they pick up the idea of the older man. Then they run with it. They take it away from the older man simply by developing it in a way that he cannot keep up with. . . . The older fellow is exceedingly fond of such a man as that. . . . Another young man comes along. He fights everything the older man says. He proves everything the older man says is wrong. . . . He thinks that is the way to get ahead. This does not make for good friends at all."[118]

HUMAN MISSIONS AND THE PROBLEM WITH PICTURES

> So many of these people would like to have the moon be something ordinary like the rocks of the earth. I am prejudiced. I would like to have the surface of the moon be something unusual—something that would tell us about the early history of the solar system 4½ billion years ago. But many of the people would be glad to have the moon be exactly like the earth. If that is the case, here is one taxpayer that does not think that it is worthwhile.
>
> HAROLD C. UREY IN A LETTER TO GEORGE DE HEVESY, DECEMBER 29, 1960

Although Urey's disappointments were in many respects personal, they were also brought on by the political nature of NASA's Cold War mission, which was to establish a strong American presence in space, not to provide an inspiring view of the universe. This became increasingly clear after President John F. Kennedy's announcement in 1961 that the United States would land a man on the Moon before the end of the decade. NASA Administrator James E. Webb had convinced Kennedy that "to be pre-eminent in space we must conduct scientific investigations on a broad front . . . in the minds of millions, dramatic space achievements have become today's symbol of tomorrow's scientific and technical supremacy."[119] Manned missions to the Moon became the US response to Soviet manned flights into space, and manned spaceflight became synonymous with scientific exploration. NASA reorganized itself around the Apollo program in November 1961, and Homer Newell now became director of the new Office of Space Science.

According to the historian David DeVorkin, Newell interpreted the new structure and priorities of the space program to mean that he had no choice but to link space science to Apollo. The robotic lunar missions that Newell had overseen now became precursors to the "real science" of having human observers on the surface of the Moon. This in turn led to a redefinition and reorganization of the priorities of existing lunar missions. Project Ranger, for example, had been designed to be a multifunctional, hard-landing lunar probe. After the announcement of Apollo, however, Ranger's scientific program was cut and the remaining probes' missions were repurposed as support for Apollo. Selecting safe and interesting landing sites now became a top priority. Any "pure science" experiments that flew onboard Ranger or the later Surveyor lunar landers were now tasked with providing engineering information for Apollo.[120]

This decision limited the pre-Apollo missions mostly to the photographic reconnaissance of possible lunar landing sites. Urey did not respond favorably to this decision. Throughout the 1960s he argued that the pictures taken of the Moon, while "very interesting," were "rather difficult to interpret." In them, he explained, "each person sees . . . exactly what he expected to find there—evidence of volcanism, of movement of dust, for fragmented material, liquid water, and so forth."[121] Some evidence of the chemical composition of the Moon had been gathered on robotic missions prior to the reorganization of the program, and it was this evidence that Urey preferred to privilege. He told his colleagues that he reserved final judgment about the Moon's origins for the day that laboratory analysis could be performed on returned lunar samples, and he encouraged them to do the same.

To Urey's chagrin, however, this reliance on photography and telescopic observation allowed a new group of scientists to rise through the ranks of NASA: the planetary geologists. These geologists would later congratulate themselves, in the words of Donald Wilhelms, on making the Moon "a world of rock" and dethroning "the physicists and other quantitatively minded scientists who once dominated space science."[122] The field geologists within the United States Geological Survey (USGS) who took up the study of the planets (and who labeled themselves "astrogeologists") carved out a prominent place for themselves within the newly forming planetary science community—a place from which they were able to reinvent their methodologies via new remote-sensing technologies.[123]

The USGS Astrogeology Branch began its existence in 1960, under the direction of the geologist Eugene Shoemaker. Although it did incorporate geochemistry, the Astrogeology Branch was strongly defined by its field component. Shoemaker and his colleagues could not do fieldwork directly on the Moon. However, using telescopic observations and NASA photographs, they attempted to produce the types of geological maps that a geologist would construct from field observations. In their previous work, many of the astrogeologists had learned techniques of photogeology, and they had become accustomed to making maps from aerial photographs. The astrogeologists began making geological maps of the Moon using the Lick Observatory's thirty-six-inch refractor and the Lowell Observatory's Clark refractor. The derivation of stratigraphic maps from photographs and remote imagery was further reinforced when NASA built the astrogeologists their own observatory, staffed by USGS personnel and used exclusively for lunar mapping. The trend continued with the lunar images returned by the robotic Ranger, Surveyor, and Lunar Orbiter missions, all of which carried cameras and returned thousands of black-and-white images via television signal.[124]

The astrogeologists tended to treat the Moon as an extension of terrestrial terrain. The Brown University stratigrapher Thomas Mutch, one of the few geologists outside the USGS to take up astrogeology during this period, wrote the first textbook of planetary geology.[125] He described the history of terrestrial geology as the exciting story of how fundamental questions had been asked and answered through the slow and persistent study of Earth.[126] That the same type of work could now be done on the Moon, and that it might even uncover a stratigraphy comparable to that found on Earth, meant that fundamental questions could be asked once again. It is in this context that Mutch introduced photogeology as the successor to traditional fieldwork. In the absence of actual ground truth, Mutch advocated the use of Earth analogues in the stratigraphic interpretation of telescopic and spacecraft images, allowing astrogeologists to bring terrestrial field experience to bear on lunar surface features. For his frontispiece, Mutch chose an Ansel Adams photograph, *Autumn Moon, the High Sierra from Glacier Point*; portraying the Moon as it might be viewed by a field geologist on a clear night, the photo highlighted one of the book's main themes—the Moon was unexplored territory, but it was connected to something familiar and mapped.

Urey clashed continuously with the astrogeologists, claiming that they were "second rate" scientists who knew little basic science and had few publications to their names. In 1961, for example, Urey wrote in a letter to Al Hibbs at JPL:

> I believe that geology is the worst training that can be given to a man for the investigation of the Moon and planets. The reason for it is this. Geology ... is largely a purely descriptive science with a very minimum training in the more mathematical aspects of chemistry and physics. It largely deals with description of rocks, sedimentary and igneous, and the gross features of the earth. I have been associated with geologists now for some 10 years and I have met a few who are good exact scientists, but the training which students get in the usual geology department is a very descriptive training.[127]

Urey explained that the processes and theories geologists were familiar with would be of no use regarding the Moon, which he believed had a very different past than Earth. Thus they would be blinded by their preconceptions and their allegiance to such theories as uniformitarianism. Urey revealed his Manhattan Project biases when he went on to suggest that space scientists should be chosen mainly from those disciplines that attracted the best and brightest scientific minds—nuclear physics, solid-state physics, and physical chemistry. Geology was not one of these disciplines; as Urey once explained to NASA associate administrator George E. Mueller, "[We] all know that geology attracts the less brilliant type of scientists."[128]

To be clear, Urey knew he could not do the work of a hard-rock geologist. With Lowenstam he had made trips to the American South and had watched his geologist colleague "spot exactly the beginning of the Eocene period in the top of the Cretaceous in the rocks which I could not recognize at all."[129] On a later vacation with his family in Colorado, Urey tried his hand at geological collecting, only to find that he could not determine what was important and what was not: "I just did not know where I was in geological collection."[130] From these experiences, Urey concluded that geology did in fact "require an enormous amount of careful personal observations," and that "a mature scientist finds it difficult to become expert in fields of this kind."[131] This type of observational work did not suit Urey, as he admitted: "I had studied geology as a student in college and had found it to be an exceedingly boring subject. I just couldn't get particularly interested in all of the rocks that were laid out for us to study at the time."[132]

Urey did recognize that geologists were essential for lunar exploration. He told Mitroff, "I've always said the geologists must be used in the interpretation of the moon. They must be used, they are the ones that know what igneous rocks are like. I don't. Other people don't. They know when they pick up a rock, that this belongs to such and such a class of rock, I don't."[133] He even insisted to NASA's Office of Manned Space Flight that sample collection to determine the origin of the Moon would require that "all astronauts be well trained hard rock geologists."[134] However, Urey felt that, as it was, the lunatics were running the asylum. Geologists were valuable as observers and collectors, not as theorists. Instead of performing the service role to which they were best suited, the geologists were actually guiding NASA's mission planners.

Urey was right to believe that he was losing his fight for the Moon. An internal NASA memo from the director of Apollo lunar exploration responding to Urey's criticisms confirmed Urey's suspicion that the astrogeologists were having a great influence on the Apollo missions: "We have turned strongly to astrogeologists for advice on site selection in the past because, with our present paucity of knowledge, the topography and stratigraphy of the lunar surface has been our key input. . . . The astrogeologists were of major assistance in [Ranger, Surveyor, Lunar Orbiter, and now Apollo] and have been responsible for a major portion of the data analysis."[135] The astrogeologists had successfully translated NASA's priorities into the development of their own niche within the agency.[136] Urey, on the other hand, had become a very distinguished thorn in NASA's side—one they preferred to ignore rather than remove.

In a 1967 argument in favor of space exploration, Urey countered the physicist Max Born's criticism that space travel, and especially the space race, held little benefit for humankind. Rather than being "a common undertaking of all peoples which would act for the reconciliation of antagonisms and the maintenance of peace," Born claimed it was "a symbol of a contest between the great powers, a weapon in the cold war, an emblem of national vanity, a demonstration of power."[137] It was not "a lightning conductor of our inborn aggressiveness and violence" that would prevent war, but "a preparation for war, a dangerous game."[138] The relatively small scientific gains from exploring the Moon and planets were not worth the illusions of superiority and the resulting increase in nationalism. Urey rallied to the defense of space science and lunar exploration, but in a very limited fashion:

> Of course if the moon escaped from the earth all we can conclude is that an insignificant planet, Earth, made a mistake, got too much angular momentum, and solved its problem by throwing off an insignificant cinder in space. If this is the case, I shall be immensely disappointed and shall feel that my attention to this subject in the last years has been wasted time and effort. But if the moon is a primitive object captured by the earth and has on its surface at least a partly very ancient surface that enables us to say something about the events during the very early history of the solar system, it will be an enormously interesting object in connection with the origin of the solar system, stars, etc. It might even give us a sample of material that was indeed the primitive material from which the solar system evolved.[139]

Put in stronger terms, Urey later told a NASA colleague, "I shall be sorry and disappointed if . . . the moon then will be an incidental object and not of fundamental importance. We can decide that it escaped from the earth and then 'to hell with it.'"[140]

As for the lunar mapping that was so important to the Apollo program and to the professional development of astrogeology, Urey was certain that lunar geology would be a dull subject: "There are those who fully expect to map the moon in great detail. But the mapping of the earth has been important because of its very active history and because men live here. . . . One cannot expect that this sort of interesting phenomenon will be extended to a cinder such as the moon."[141] When he spoke before the public, he often cautioned his audiences that the Moon might in fact be disappointing, but that he hoped it would be "interesting" enough to make the $20 billion Apollo program "worthwhile."[142] Urey's gamble was that the Moon would be of fundamental importance. As he told Congress, "I think that it has no real interest to us except as a way of understanding the origin of the solar system, and I hope very much that it will be important in that connection."[143]

The problem of unlocking the secrets of the solar system must have taken on greater significance for Urey in the 1960s. This was a difficult time for the aging scientist. The decade began optimistically—Urey was a member of then Senator John F. Kennedy's Committee on Science and Technology of the Democratic Advisory Council, corresponded with the candidate, and met with him personally when the campaign visited San Diego. After Kennedy's inauguration, Urey wrote in an inscription to be presented to the new president:

> Science and Technology have completely changed the world in my lifetime. Most of the urgent political problems of these years have been produced very directly by these changes. Never in my life have I been so thrilled by the inauguration of a president as I have been by that of President Kennedy. We have a young, daring and fearless president who welcomes the real challenge which these changes present. He has started his administration work effectively. The problems, both domestic and foreign, are so great that he may well be recorded in history as one of our greatest presidents. We all wish him great success and hope that he earns that niche in the history of our country.[144]

In 1962, Urey celebrated his 69th birthday at a White House dinner. When, less than one year later, Kennedy was assassinated, Urey was devastated. He wrote to Jacqueline Kennedy, "It seems to me that never in my life have I felt so badly about such a fearful thing as that which happened to your husband and, of course, to you and your children. To me, President Kennedy was the most wonderful president of my lifetime." For the remainder of his career, Urey kept an autographed photo of Kennedy in "an honored spot" in his office.[145]

Urey was similarly devastated by the assassination of fellow Nobel laureate, the Reverend Martin Luther King Jr. Urey admired King's fight for civil rights and, of course, his religious courage in standing up to violence and oppression. Less than one week after King was killed, Urey began a letter-writing campaign to his fellow Nobelists, raising funds to memorialize King with a donation to the Southern Christian Leadership Foundation. In the statement he crafted for the group, Urey wrote:

> We grieve at the silencing of the eloquent humanitarian voice of Martin Luther King, Jr. We share what he has called an audacious faith in the future of mankind. We also have "the audacity to believe that peoples everywhere can have three meals a day for their bodies, education and culture for their minds, and dignity, equality, and freedom for their spirits." We shall enshrine him in our memories as one who attempted to increase the brotherhood of men and to "have conferred the greatest benefit on mankind."[146]

The martyrdom of both of these high-profile, heroic figures no doubt took a toll on Urey's optimism. Would a "new prophet" even survive to bring peace to humankind?

INCONSTANT MOON

With so much institutional resistance to his ideas and his inability to exert influence on NASA in the planning of lunar exploration, it is no wonder that Urey never managed to revise *The Planets*, let alone to produce a popular work on the origin of the solar system for his prophet. Unceremoniously, Urey wrote to Anshen in 1967 to tell her. Echoing what he had written to Yale University Press, he lamented to Anshen, "I have not thought seriously of writing a book at all, but if I were going to I would try to revise my book on the planets. This of course would have to be published by Yale University Press and would not give you any help."[147] Urey was beaten. The only hope he held out was that the Moon might yet prove to be as important as he hypothesized, and that he might then get to resurrect his lunar theories. In the end, however, the Moon would not cooperate with Urey's expectations.

The first lunar rocks brought back by Apollo 11 were distributed in 1969 to the Apollo research teams that performed various forms of age measurements and mineral analysis. In some ways, Urey and his fellow cold Moon advocates were vindicated by the findings from the Apollo 11 rocks and those that followed—all the measurements agreed that the rocks were billions of years old. Some rocks from later Apollo missions proved to be 4.5 billion years old, almost as old as the solar system itself, just as Urey had predicted. This prompted some lunar scientists to tell the *Washington Post*, "You know that Harold is the grandfather of us all. . . . Uncle Harold is the real modern father of lunar science. . . . Don't let anyone tell you he isn't as sharp as he ever was."[148]

Still, the Moon turned out not to be anything that likely predated the formation of the planets. It lacked many of the materials that Urey predicted a primary object should contain. It was completely depleted of volatile elements. The hot Moon advocates were also vindicated. Although the youngest rocks on the Moon were roughly 3.8 billion years old (meaning that it had been geologically inactive for most of its history), it had experienced significant melting in its early history.

Urey was wrong about the uniqueness of the Moon. It did not represent a fossilized remnant of the early solar system. While the Moon could still be considered a Rosetta stone of sorts—it did record the history of a period of bombardment in the inner solar system that had been erased from Earth's surface—it was not the geochemical Rosetta stone that Urey hoped it would be. As Newell explained:

> Harold Urey has said many times that he expected the Moon would be most interesting—that he hoped the Moon would turn out to be an interesting object—that he feared the Moon would not turn out to be interesting after all. Now that Ranger, Surveyor, Lunar Orbiter, [the Soviet Luna and Lunokhod spacecraft], and Apollo have flown, providing samples of lunar material for laboratory analysis and vast quantities of photographic and instrumental data on the physical and chemical characteristics of the Moon, many of the Moon's secrets are being exposed to view and scrutiny. The Moon is very old, as old as the solar system, as Harold thought and hoped it would be. It is not, however, in its primitive state, having clearly undergone considerable evolution in the first 1500 million years after its formation. . . . Nevertheless, many parts of the lunar surface are appreciably older than most of the Earth's surface, and the Moon may yet prove to be the Rosetta stone of planetary origins, as it was dubbed in the early 1960's. In any event, it is an important and illuminating link with the remote past, and its study will have much to reveal about how the Moon and planets formed and evolved. I hope that Harold Urey has decided that the Moon is after all a most interesting object, that has been worthy of all the competence and insight that he has brought to bear upon the study of it.[149]

But was Urey happy with this Moon?

Despite all the conflicts and unhappiness that NASA's lunar program had brought him, Urey had enjoyed the adventure. However, he did write in 1976, "Yes, I think the moon has been quite a disappointment to me. I thought it would tell us something unique about the solar system. However, it seems to be an incidental object of some kind with no theory for its origin that is generally accepted."[150] Indeed, no generally accepted theory of the Moon's origins would emerge until 1981, just after Urey's death.[151] Not only was a great inspirational narrative no longer possible, even the story of how science had unlocked the Moon's secrets could not yet be told.

EPILOGUE

A Life in Science

Just before leaving Chicago for La Jolla in 1958, Harold Urey had his portrait painted by the Norwegian-born artist Christian Abrahamsen. Upon viewing the portrait some years later, Harold D. Lasswell, a political scientist and fellow participant in the Conference on Science, Philosophy, and Religion, remarked,

> The artist's conception of Urey is strong enough to indent the wall and dissolve the ceiling. Urey sits with the rock-heavy power of a tomb figure from the valley of the Nile. He is a Pharaoh of the mind—by virtue of achievement. Urey's lively eyes see through the physical confines of the studio to the limits of our solar system, and into the galaxies beyond.

Twenty years after the painting was first shown, Urey felt like a very different man. The artist wrote to Urey in 1979, describing a photograph of him in his laboratory, under which someone had written, "Retirement? Not for Urey, He's Too Busy" (a joke that circulated around Chicago when the retirement-age chemist left for his new post in California). Urey wrote back, "Yes, I felt like the caption under the picture for many, many years. However, now I take joy in sitting on my beautiful patio and viewing my flower/orchid garden."[1]

Urey's letter to Abrahamson painted a pretty picture of retirement. In fact, however, his body had betrayed him. By 1975, at the age of 82, he had stopped traveling. He wrote in 1977 in his annual birthday letter to Raymond T. Birge, "I am feeling my old age every day. My hands quiver, I wobble when I walk, my eyesight is bad and I can't remember things so that I cannot keep up with the literature, hence I can do no scientific work at all."[2] The power of concentration that had defined him for much of his life was gone. The symptoms he described were related to his two main ailments, Parkinson's disease and macular degeneration. One newspaper headline reported that the once great man lived "in a failing body."[3]

Over the next year, Urey retreated from science completely. He stayed home and spent much of his time sitting in his garden, under Frieda's care and with the help of a secretary who came to the house to keep up his correspondence. It was the first time in his adult life that he was not working vigorously at something, and it was a difficult transition. The "lively eyes" that in his portrait could see "the limits of the solar system" were now blurred and confined to the house; when the Voyager spacecraft sent back close-up images of Jupiter, he was too ill to witness their unveiling at the JPL mission control room. Instead, Frieda kept him company as he "sat glued to the TV for every viewing of the flights."[4]

He was particularly interested in Jupiter's moons. "Some time ago, I suggested there were many moons in the early solar system, and then I concluded it was not right," he wrote to Birge, "but, when I looked at the 4 moons, it seemed to me that they must have been captured by Jupiter, and that means a considerable number of moons must have been around in the early solar system."[5] He lamented that he did not have the energy to go out on one last scientific limb. As he reported to Birge just a year later, his eyesight was so bad that he could read only newspaper headlines, nothing smaller.

In these final years, the topics that had once energized Urey's life and work now elicited mixed emotions. Science had fulfilled many of its promises: "We have lived through very interesting times in science. Think of the things that have happened in this century. Almost the entire development of radioactivity has happened in this century. The whole development of isotopes also." But at the same time, he was plagued by the uncertainties that those great developments had unleashed. In his next birthday letter, Urey asked Birge, "Do you think people will be here a million years from now, or even a century from now? . . . I wonder if we have not lived through an exceedingly interesting time and just before a very dreadful time when problems and disaster will plague men on the earth."[6]

In one of his last public speeches, he told the graduating seniors of McGill University that, although he and his colleagues had had "fifty-three years of comparative professional success and great prosperity," he did not feel confident that they would leave the world better than they had found it. There had been triumphs, but "one great cloud" still hung over their heads. Already at the end of his own career, all he could do was wish the younger generation luck at clearing that cloud away.[7] While he had successfully transformed his research program into one focused on the earth and planetary sciences, and had managed to become the "grandfather" of lunar exploration, he had failed to find a way to use this research to provide the world with what he felt it most needed—inspiration.

On January 5, 1981, Harold C. Urey died. He did not know if he had left the world better or worse for his efforts. This was the one final calculation that eluded him. His body was cremated in La Jolla, California, and then buried in DeKalb County, Indiana. Other planetary scientists would have their ashes sent to the Moon or into space, but Urey insisted on being sent home, to where his journey began. "The cemetery is small and modest, and the tombstone is plain. It's on high ground . . . far from large towns or highways, surrounded by farm fields and wooded land." His Corunna obituary noted that the Nobel Prize winner laid to rest in the Fairfield cemetery had "helped preside over the birth of the atomic age and made discoveries that lie at the foundation of modern science."[8] Eleven years later, Frieda joined him in their final resting place.

. . . . .

What should we make of Urey's life in science? He did, as his obituary suggested, help usher in the atomic age. He initiated a generation of American scientists into the world of quantum physics and chemistry, translating the work coming out of Copenhagen to his peers at home. His discovery of deuterium and heavy water contributed greatly to the development of nuclear reactors and the hydrogen bomb—although he himself came to fear nuclear power plants and a possible "China syndrome" that might result from a reactor core meltdown. He likely would have preferred to be remembered for his work in the development of isotope geochemistry. In this area, he produced a cohort of younger scientists who brought his vision to life at universities beyond Chicago's Institute for Nuclear Studies. Samuel Epstein became a leader in the field, based at Harrison Brown's new facilities at Caltech, even as Brown himself stepped away from science and into the realm of policy. Cesare Emiliani brought mass spectrometry to the University of Miami's Institute of Marine Sciences and continued a strong research program on carbon and oxygen isotopes in fossilized marine organisms. He helped organize the Joint Oceanographic Institutions for Deep Earth Sampling and contributed to work that has reconstructed the past climate of Earth and informed our present understanding of climate change. Stanley Miller, who moved with Urey to the University of California, San Diego, continued his research on the origin of life. Another of Urey's Chicago students, Harmon Craig, brought isotope geochemistry to the Scripps Institution of Oceanography. Among his many achievements, Craig studied the interactions of the deep earth and oceans, including deep-sea hydrothermal vents—another possible venue where life may have originated.

In 1989, the International Society for the Study of the Origin of Life, on Miller's initiative as president of the society, instituted a Urey Medal for contributions to the field of origin of life studies, to be awarded in alternate years to the society's already existing Oparin Medal. In addition to recognizing Urey for his contributions to the field, the new medal afforded the society a solution to a late–Cold War dilemma. The late 1960s had brought revelations by Zhores Medvedev and David Joravsky of Oparin's and other Russian scientists' alliances with Trofim Lysenko.[9] As one disgruntled member wrote to the society, citing both Medvedev and Joravsky's work:

> More than 3 years ago, the street address of the Soviet Academy of Sciences has been changed to Vavilov str 32, Moscow. . . . Vavilov is widely regarded as a martyr of science. Vavilov was hounded to death by Lysenko . . . and died of malnutrition in the Saratov prison. Oparin "joined the Lysenkoite movement, the only really eminent biologist to do so. From 1948 to 1955, he was in charge of Lysenkoite firing and hiring within the (USSR) Academy of Science. . . . He altered his speculations on the origin of life to suit the Lysenkoite creed. . . ." . . . Now that the winds of perestroika are blowing, it is time

> that ISSOL revise the name of its major award. In my opinion, the change would be welcomed at Vavilov str. 32, Moscow.[10]

Those honorees who wished to refuse the Oparin Medal on political grounds were permitted to accept the Urey Medal the following year.[11] Even from beyond the grave, Urey's science couldn't escape the Cold War's gravitational pull.

Beyond geo- and cosmochemistry, Urey also had a great influence on the lives of at least a few women scientists. Mildred Cohn, Urey's first Jewish woman graduate student in the 1930s, became an accomplished biochemist, well known for her work on the chemistry of metabolic processes, and a model for women and minorities in the sciences. By that same token, Urey's eldest daughter, Elizabeth Baranger, became a well-regarded physicist in her own right, and eventually vice provost at the University of Pittsburgh.

There was also Toshiko "Tosh" Mayeda, the Japanese American woman Urey hired in 1950 as a lab assistant. She had spent World War II in California's Tule Lake internment camp before receiving an undergraduate degree in chemistry from Wilbur Wright College. In Urey's lab she assisted Epstein and Urey, coauthoring papers with both men, and received the equivalent of a doctoral education in physical chemistry and mass spectrometry. When Urey left for California, Mayeda stayed at the lab and collaborated with its next occupant, the chemist Robert Clayton. The two used the McKinney-Nier spectrometer Urey left behind for the next forty-five years and published extensively on oxygen isotope ratios in the solar system. They studied meteorites, and when the Apollo 11 lunar samples were distributed to scientists, Mayeda and Clayton received lunar soil for analysis.[12]

Urey attempted to be a voice of reason in the Cold War. His solution to the danger posed by nuclear warfare and the failure of nations to resolve their differences was to promote religion as a source of morality. In doing this he drew from his own upbringing within the Church of the Brethren, even if he never explicitly claimed this heritage. His emphasis on the Bible and Judeo-Christian tradition—particularly the Ten Commandments and the Sermon on the Mount—as civilizing forces that had made human progress possible would have found welcome audiences within the churches his father and stepfather had served. Growing up among the Brethren was also the root of Urey's understanding of Christianity as a communal practice aimed at self-sublimation, expressed in outward forms and enforced by powerful taboos. This lived experience of community seems to have been at work in his characterization of the ideal scientist and scientific community, even as he often presented it in the language of cosmopolitanism.

Urey also attempted to reinvent himself as a cosmic storyteller, only to be met with the resistance of a great scientific and professional bureaucracy. He had managed to move with the major trends during his lifetime, had risen from farm boy to scientific star, had participated in many of the great scientific achievements of the twentieth century, and had even managed to create new fields on the boundaries of disciplines. But the world Urey inhabited after the war was not a world with which he was familiar; it certainly was not the world that had celebrated him in the 1930s. The scale of research after World War II did not suit his personality. The introduction of the contract research system and "big science" benefited his research program but also transformed the size and structure of science in ways that diminished his authority. This is obvious in his demoralizing experiences as the director of Columbia's SAM Lab and as a participant in NASA's lunar exploration program. Science and technology were now valuable resources for the national defense, tools of foreign policy, and forms of statecraft. Like his friends J. Robert Oppenheimer and Edward U. Condon, Urey had entered the Cold War with a particular vision of the scientist as a special type of citizen with a special type of expertise—not a politician in his or her own right, but a valuable adviser and mediator in times of crisis. They had seen their attempts to intervene in Cold War politics met with hostility and claims of disloyalty. The politicians were not interested in their authority, only their service.

Urey's attempt to reconcile science with religion, even if he failed to accomplish this, is a strong reminder that neither science nor religion is an abstract concept or philosophy. They are both social enterprises and lived experiences, and their relationship is determined both communally and individually by those who participate in them. Urey did not think of science and religion in a vacuum, but drew on the several intellectual and social traditions within which he had lived, and the various worldviews he had held.

Urey's early faith in science as a positive force, and in the progressive tendencies of society, was contextual. As a young man whose own rise to fame and fortune paralleled the growth in prestige of American science, it was natural that Urey saw science as a noble profession that could bring abundance and equality to all corners of the world. But by the end of the Second World War, Urey was not so certain of science's ability to control itself, nor of society's ability to rein it in. Though scientific research remained his primary passion, the progress of science seemed to trend more toward materialism and destruction. The shared value system Urey took for granted as the basis for a progressive society in the 1920s and '30s seemed fractured in the Cold War. Scientists were deeply embedded in the bureaucratic structure of the state.

While his ideas for a reconciliation of science and religion may now seem naive or even incoherent, one must admit that he was ahead of his time in at least one respect. Urey realized that science was a human activity, practiced in a social context. The values of society were the values of science—the knowledge and technology produced by scientists and engineers would reflect those values. Furthermore, he also realized that the influence went two ways. Science, its practitioners and its products, could push society toward materialism or toward inspiration. Society, with its national interests, political processes, and economic drives, could push science in various directions. Neither was "pure."

It is safe to say, though, that science always remained special in Urey's mind. The persona of scientist gained him entry into the most elite institutions in America. His accomplishments brought him more success than he could have imagined as a boy. However, Urey wanted more, not for himself but for society. He wanted science to save the world. When it failed to do so, when it became ever more bureaucratized and entwined in issues of geopolitics and national security, he could not help but be disappointed. Does this make Urey a tragic figure? Perhaps not. But it left him at least one step short of what he would have considered heroic.

Acknowledgments

This book is wholly my own work, but it could not have been completed without the help of many individuals and institutions. In the history of science community, I have benefited from the assistance of two primary mentors. The first is Dr. Naomi Oreskes of Harvard University, who supported every turn that this research project took and provided invaluable feedback and encouragement. I first approached Naomi because of an interest in the history of geology and its application in planetary science. This ultimately turned, at least temporarily, to an examination of Harold C. Urey's role in the development of isotope geochemistry, and then to the biographical treatment of Urey contained herein. Naomi encouraged this trajectory, agreeing that Urey was an interesting case study in Cold War science—and neither of us could resist the religious history I uncovered in his papers that had been omitted from his official profiles. While other scholars (and even a few academic press editors) advised against writing a biography as a first book, Naomi insisted that I simply make it a relevant biography that spoke to major themes in the field. And so, here we are. Fortunately, the University of Chicago Press, my editor Karen Darling, and the Synthesis editorial board agree that we made the right choice.

I also am indebted to Dr. Jane Maienschein and the Arizona State University Center for Biology and Society. Jane introduced me to the history of science when I hardly knew that such a field existed. Her work on the history of biology and medicine—and the questions she has tackled about the meaning of science and the people who practiced it—inspired me to take on projects of my own. It was with Jane's help that I first started working on topics related to planetary science and exploration with Dr. Phil Christensen of ASU's Mars Space Flight Facility.

It was Robert Westman who started me thinking about science and religion as I made my first foray into the Urey papers. I went into the archives searching for clues into Urey's religious life, knowing only what Daniel Kevles had written in *The Physicists*: that his father had been a minister and that he subscribed to a "secular faith" of his own. When I discovered Urey's public speeches from the 1950s and read his plea that humankind needed a new prophet to bring together the moral teachings of the great religions with the dramatic (but not miraculous) view of the universe provided by science, this finding became the kernel for this book.

The UC San Diego Science Studies Program and Department of History provided an incredible intellectual atmosphere within which to work. Martha Lampland, Charles Thorpe, Andrew Lakoff, and Steven Epstein helped form my understanding of the social, cultural, and political nature of science. Robert Edelman and Frank Biess advised me in my study of Russian and Soviet history and gave me a much better grasp of the use of social and cultural methods in history than I would otherwise possess. UCSD also provided financial support for the original research presented here. Some of this money was provided to the Science Studies Program through a private endowment from the estate of Rik and Flo Henrikson. The UC Humanities Network also supported me as one of its predoctoral Humanities Research Fellows. UCSD's Kenneth and Dorothy Hill Fellowship funded the final year of my research in the Urey papers. The support of a National Science Foundation grant afforded me the necessary research materials, travel, and time spent at several relevant archives. For one very productive year during my research, I was a Haas Fellow at what is now the Science History Institute in Philadelphia, Pennsylvania.

The University of Southern California and Harvard University provided me with research support in the latter years of this project. I wish in particular to thank Peter Westwick for bringing me to USC under the aegis of the USC–Huntington Library Aerospace History Project and for encouraging my continued study of Cold War science. At Harvard I had the luxury of discussing the biographical genre with Janet Browne and soliciting her advice as I continued my revisions.

I now occupy the position of curator of planetary science in the Space History Department of the Smithsonian's National Air and Space Museum. I want to thank all my colleagues in the department. In particular, Michael Neufeld has helped me through the final stages of writing and revising this book. Mike read every word of the book (including the many words no longer in it, thanks to him) and provided expert advice on structure and organization. David DeVorkin, Layne Karafantis, Margaret Weitekamp, Jennifer LeVasseur, Martin Collins, and Roger Launius also read portions of this book and provided very helpful suggestions as drafts progressed. None of this would have been possible without the support of my department chair, Valerie Neal. The high standards of scholarship maintained by my colleagues and the museum's leadership allow the museum not only to contribute to the Smithsonian's mission but to remain a world-renowned center for the history of aviation, spaceflight, and the associated science and technology.

It would be difficult to list here all the colleagues who have helped me in ways large and small. My views on science studies and the history of science are the result of countless discussions over the last several years—most frequently with Elena Aronova, Luis Campos, Philip Christensen, James Collins, Matthew Crawford, Richard Creath, Monica Hoffman, April Huff, B. Harun Küçük, Manfred Laubichler, Brian Lindseth, Eric Martin, Patrick McCray, Minakshi Menon, Cynthia Schairer, James Strick, and Alex Wellerstein.

There are also many librarians and archivists to thank. My primary archive for this research project was the Mandeville Special Collections Library (MSCL) at UCSD, where the papers of Urey and his La Jolla colleagues are deposited. Lynda Claassen, the department head of MSCL, and her staff helped me navigate the extensive collection. Urey's career was spent at multiple institutions on the East and West Coasts. In New York I visited Columbia University, where I had the assistance of Gerald Cloud in the Rare Book and Manuscript Library and Courtney Smith in the Oral History Research Office. Thank you to Harriet Zuckerman for granting me permission to use the oral history interview that she conducted with Urey. While in New York I also visited the Jewish Theological Seminary of America, where Sarah Diamant assisted me with the papers of the Conference on Science, Philosophy, and Religion. At the University of Chicago's Special Collections Research Center, directed by Alice Schreyer, I was happy to have the knowledgeable guidance of Julia Gardner.

Discussions with two of Urey's younger colleagues at the University of Chicago's Institute for Nuclear Studies (known today as the Enrico Fermi Institute), Roger Hildebrand and Robert N. Clayton, helped me understand the structure and atmosphere of the institute during its early years. It was Clayton who first emphasized to me the importance of mass spectrometer development in the immediate postwar years, and who encouraged me to take a closer look at Samuel Epstein's papers at Caltech. At the Caltech Archives, headed by Shelley Erwin, I was assisted primarily by Loma Karklins. In Washington, DC, I spent time at NASA's History Division with its then director Steven J. Dick and his knowledgeable staff of historians and archivists. In the Ava Helen and Linus Pauling Papers, held at Oregon State University, I enjoyed the assistance of the head of Special Collections, Cliff Mead, and faculty research assistant Chris Petersen. The American Institute of Physics Center for the History of Physics was very helpful in providing Urey's oral history interview, conducted by John Heilbron, as well as transcripts of Ian I. Mitroff's interviews with Urey during NASA's Apollo program. I have enjoyed talking physical chemistry with AIP's Greg Good, director of the Center for the History of Physics, and Charles Day, editor of *Physics Today*.

During my year at the Science History Institute's Beckman Center for the History of Chemistry, I was able to make use of the institute's Othmer Library, Oral History Program, and archives. I wish to thank the staff, especially Ronald Brashear, Anke Timmerman, James Voelkel, David Caruso, Hilary Domush, Sarah Hunter, Andrew Mangravite, Elsa Atson, Ashley Augustyniak, Hyungsub Choi, and Jody Roberts. I would also like to thank the invigorating community of fellows—including the fellows of the Philadelphia Area Center for the History of Science (now known as the Consortium for History of Science, Technology and Medicine). Especially deserving of thanks for their help in conceptualizing various parts of this project are Benjamin Gross, Evan Ragland, Annalisa Salonius, and Carin Berkowitz. While in Philadelphia I was able to get to know one of Urey's former students, the chemist Mildred Cohn. Professor Cohn unfortunately passed away before she could complete her own biography of Harold Urey, and before she could read this book.

I was fortunate to have had the friendly cooperation of two of Urey's children, Elizabeth Urey Baranger and John Urey, and one of Urey's nieces, Paulette Cullen. I hope that the Harold Urey I have constructed bears at least some resemblance to the man they knew and loved. For information about Urey's younger years, I must thank Janet Rhodes of the Walkerton (Indiana) Area Historical Society, Amy Weiss of the University of Montana's Archives and Special Collections, and Amy Rupert of the Rensselaer Polytechnic Institute Archives and Special Collections. For his assistance in tracking down genealogical source material in the Waterloo, Indiana, public library, I must thank William M. Ellerman II.

Finally, and most personally, my most sincere and heartfelt thanks go to my family. My parents, Richard and Mary Shindell. Their support made everything possible. And thanks to my wife, Jeannette Shindell, who met me before the business of writing a first book began. I know that she is as happy as I to see it completed at last.

Notes

INTRODUCTION

1. This narrative is compiled from two separate and slightly differing accounts Urey provided in Stanley Miller, "Harold Urey—Biographical Memoirs (Period 1923–1939)" (typescript, n.d.), p. 10, box 191, folder 9, SM; and Hall and Urey, "As I See It," 48.

2. Urey, Brickwedde, and Murphy, "A Hydrogen Isotope of Mass 2"; Urey, Brickwedde, and Murphy, "A Hydrogen Isotope of Mass 2 and Its Concentration," 1.

3. Rather than concede that the definition of isotope should be changed, Soddy in fact argued that deuterium should be considered an element on the grounds that it violated his definition. See Eyring, "The New Point of View in Chemistry."

4. E. F. A., review of *Inorganic Chemistry*, 431.

5. The 1930 US Census lists the family as living at 29 Claremont Ave in Manhattan, lists Margaret Strickland as a "roomer," and indicates that they paid $125 in monthly rent.

6. The 1940 Census lists the family as living at 355 Highland Avenue in Palisades Park, New Jersey, lists Sherman as a servant and indicates that the Urey family paid her an annual salary of $400.

7. Lord Rutherford et al., "Discussion on Heavy Hydrogen: Opening Address."

8. *New York Times*, "Our Place in Science."

9. *New York Times*.

10. On the effects of World War I on physics and chemistry, see Nye, *Before Big Science*. The increase in the numbers of American scientists and institutions of science after World War I (and the connection of this new scientific community to Bohr's institute) is presented in Kevles, *The Physicists*, chap. 14, "A New Center of Physics."

11. Although Richards received his bachelor's and his doctorate from American schools (Haverford and Harvard), he had been raised in England and France, and, like most chemists of his generation, had also studied in Germany prior to taking up a university appointment in the United States. Langmuir studied in Göttingen under the direction of Walther Nernst. By contrast, Urey had studied at the University of Montana and the University of California.

12. Craig, Miller, and Wasserburg, *Isotopic and Cosmic Chemistry*, iii.

13. *New York Times*, "Our Place in Science."

14. Compton, "Science Still Holds Great Promise."

15. Stevens, "Harold Urey: A Genius Lives in Failing Body."

16. Harold C. Urey, "Autobiography, 1970" (unpublished manuscript), p. 1, box 1, folder 5, HCU.

17. See Milam and Nye, *Scientific Masculinities*.

18. The Nobel biography of Urey was originally written for the award presentation in 1934, and was later published in Nobel Foundation, *Nobel Lectures, Chemistry 1922–1941*. A draft version of this brief biography also exists: [Harold C. Urey], "Biography Prepared for Nobel Prize Committee" (typescript, n.d.), box 1, folder 9, HCU.

19. C. F. Bowman and Bowman, *Brethren Society*, 23–94.

20. See Cohen et al., "Harold Clayton Urey"; and Arnold, Bigeleisen, and Hutchison, "Harold Clayton Urey."

21. Urey's own discussion of his Brethren roots in his autobiography is included primarily to illustrate the roots of his pacifism. A 1970 biographical introduction of Urey prepared by his colleague Joseph Mayer reproduces this point. See Mayer, "[Biography of Harold C. Urey]" (typescript, 1970), box 1, folder 11, HCU.

22. Harold C. Urey to Louis Finkelstein, August 11, 1949, box 44, folder 11, HCU.

23. Harold C. Urey, "The Intellectual Revolution [Revision]" (typescript, 1956), p. 18, box 141, folder 12, HCU.

24. Urey first introduced the "new prophet" idea in 1959. See Harold C. Urey, "Science and Society—Cooper Union Conference" (typescript, November 2, 1959), p. 13, box 141, folder 23, HCU.

25. Harold C. Urey, "Religion Faces the Atomic Age" (typescript, February 3, 1958), p. 4, box 141, folder 15, HCU.

26. R. C. Hall, *Lunar Impact*, chap. 1, "The Origins of Ranger."

27. Kevles collected this interview for his doctoral dissertation, "The Study of Physics in America, 1865–1916," which later became the monograph *The Physicists: The History of a Scientific Community in Modern America*. Kevles's is without a doubt one of the most influential historical treatments of American physics.

28. See Urey, interview by Zuckerman. This interview was collected as part of Zuckerman's doctoral project, "The Nobel Laureates in the United States," which later became her monograph, *Scientific Elite*.

29. See Urey, interview by Heilbron.

30. See Groueff, *Manhattan Project*.

31. Mitroff conducted what he referred to as "psychological surveys" of the Apollo lunar scientists which he then coded and used in his publications "Norms and Counter-norms in a Select Group of the Apollo Moon Scientists," "On Evaluating the Scientific Contribution of the Apollo Moon Missions via Information Theory," and *The Subjective Side of Science*.

32. Urey had been a participant in Roe's study of scientists, although her subjects in her book *The Making of a Scientist* were treated with anonymity.

33. Urey, interview by Zuckerman, 43.

34. Kevles, *The Physicists*, 225.

35. James R. Arnold, "Harold C. Urey" (typescript, June 1, 1981), p. 1, box 191, folder 9, SM.

36. See Dennis, "Historiography of Science: An American Approach"; Greenberg, *The Politics of Pure Science*; Greenberg, *Science, Money, and Politics*; Hershberg, *James B. Conant*; Hollinger, "The Defense of Democracy"; Hollinger, "Science as a Weapon"; Hollinger, *Science, Jews, and Secular Culture*; Needell, *Science, Cold War and the American State*; Westman, "Two Cultures or One?"

37. See Krige and Pestre, *Science in the Twentieth Century*; Leslie, *The Cold War and American Science*; Lowen, *Creating the Cold War University*; Mukerji, *A Fragile Power*; Simpson, *Universities and Empire*; Thorpe, *Oppenheimer*; J. Wang, *American Science in an Age of Anxiety*; J. Wang, *In Sputnik's Shadow*; Westwick, *The National Labs*.

38. For notable examples, see Stanley, *Practical Mystic*; Davis, "Prophet of Science," parts 1–3; Davis, "Robert Andrews Millikan"; and Rupke, *Eminent Lives in Twentieth-Century Science & Religion*.

39. Banner, "Biography as History," 580.

40. Gordin, *A Well-Ordered Thing*, 239.

41. Nasaw, "Introduction to AHR Roundtable," 578–79.

42. In 2006, the journal *Isis* devoted their "Focus" section to the topic of "Biography in the History of Science," which included J. L. Richards, "Introduction: Fragmented Lives"; Terrall, "Biography as Cultural History of Science"; Porter, "Is the Life of the Scientist a Scientific Unit?"; and Nye, "Scientific Biography: History of Science by Another Means?" Much of the discussion surrounding science and biography in these recent publications can be traced back to Shortland and Yeo, *Telling Lives in Science*.

43. Rupke, *Eminent Lives in Twentieth-Century Science & Religion*, 26.

44. Livingston quoted in Rupke, 27.

45. Historian Martin Rudwick quoted in Brooke, "Religious Belief and the Content of the Sciences," 5.

46. See Brooke, Osler, and van der Meer, *Science in Theistic Contexts*; Brooke,

Science and Religion; Lindberg and Numbers, *God and Nature*; Lindberg and Numbers, *When Science & Christianity Meet*.

47. Brooke, *Science and Religion*, 5.

48. Brooke, 6.

CHAPTER ONE

1. Thomas, "Harold C. Urey," 219; *Waterloo Press*, "Thirteen Graduates."

2. Harold C. Urey, "Autobiography, 1970" (unpublished manuscript), p. 1, box 1, folder 5, HCU.

3. Martha Cullen, "My Life, 1898–1982" (La Verne, CA, n.d.), 5, MCU.

4. Harold C. Urey to *Kendallville News-Sun*, April 5, 1963, box 51, folder 14, HCU; Thomas, "Harold C. Urey," 219.

5. Mennonite and Amish numbers from 1770 are taken from Lehman and Nolt, *Mennonites, Amish, and the American Civil War*, 9; Edwards, *History of the American Baptists*; and C. F. Bowman and Bowman, *Brethren Society*, 16.

6. Untitled note, n.d., box 18, folder 10, SM.

7. By contrast, an 1880 plot map of Fairfield Center included within a family history shows that the Reinoehls owned over 150 acres, including a portion of the millpond. MCU.

8. Cullen, "My Life," MCU.

9. Harold C. Urey, "Evolution vs. Miraculous Creation," n.d., p. 11, box 144, folder 25, HCU.

10. These descriptions are taken from the text of the 1877 annual meeting, reproduced in Sappington, *The Brethren in Industrial America*, 100.

11. Harold C. Urey to Reverend Arthur Morris, August 7, 1974, box 59, folder 12, HCU. These recollections where Urey spoke of his boyhood are the same recollections he tended to reference in his 1950s public speeches about the importance of applying science and technology within the context of the traditional moral teachings of the Judeo-Christian religions.

12. Charles Yearout quoted in Hogan, "Intellectual Impact of the Twentieth Century," 85.

13. Quoted in Hogan, 85, 86.

14. Urey, "Autobiography, 1970," 1, HCU.

15. P. E. Reinoehl, *History of the Fairfield Cemetery*, 164.

16. Eshelman et al., *History of the Church of the Brethren*, 30.

17. Urey, interview by Groueff.

18. Eshelman et al., *History of the Church of the Brethren*, 30.

19. Waters, "A Little Known Boyhood from the Past," 61.

20. Waters, 61.

21. Samuel C. Urey to John Urey, July 1, 1897, MCU.

22. *Los Angeles Herald*, "A Bad Case," 10.

23. *Los Angeles Herald* "A Bad Case," 10.

24. Reinoehl, *History of the Fairfield Cemetery*, 164–65. This again is the family's version of events. It is possible that Samuel had little choice in the matter of returning home to Indiana.

25. Obituary by J. H. Elson, MCU.

26. Reinoehl and Phillips, *Ancestors and Descendants of Solomon and Martha Reinoehl*, 57.

27. Waters, "A Little Known Boyhood from the Past," 61.

28. Cullen, "My Life," 2, MCU.

29. Urey, "Significance of the Hydrogen Isotopes," 803.

30. Dove, "Cultural Changes in the Church of the Brethren," 209.

31. Dove, 209.

32. Dove, 213.

33. P. E. Reinoehl and Phillips, *Ancestors and Descendants of Solomon and Martha Reinoehl*, 11.

34. Urey, "Autobiography, 1970," 2, HCU.

35. Marguerite Cramer knew Urey well and, along with Eloise Redmond, was one of the few childhood friends with whom Urey kept in touch in his later years. Her memory of his having attended an Amish school may represent conflation on the part of the interviewee between the Brethren and the Amish (an easy mistake to make if Urey dressed in traditional Brethren clothing during these years), or it may be that the grade school Urey attended was in fact Amish in orientation. Cramer's memories of Urey are included in Housholder, "Kendallville Graduate Worked on Manhattan Project."

36. Harold C. Urey to Eloise Redmond, December 1, 1976, box 77, folder 30, HCU.

37. C. F. Bowman and Bowman, *Brethren Society*, 202.

38. Urey, "Unpublished Autobiography," 3.

39. Urey, 2.

40. Urey, 2. The freethought movement and Ingersoll are described in Jacoby, *Freethinkers*.

41. A prime example of this position would be Ingersoll, *About the Holy Bible*. This was only one of many slim 25-cent volumes produced by Ingersoll during his career as a public speaker.

42. Ingersoll, 12.

43. Ingersoll, 12.

44. Ingersoll, 12, 14.

45. Ingersoll, 18.

46. Thomas, "Harold C. Urey," 219.

47. Cullen, "My Life," 7, MCU.

48. Alva's travails are chronicled in the pages of the local newspaper, the *Big*

Timber Pioneer, in which he posted his intent to settle his homestead as well as his sale advertisement. In 1920 the Oregon Mortgage Company began foreclosure suits on eighteen settlers with outstanding debts, including Alva, who was by that time long gone. The notice of sale was printed in the *Pioneer* on March 16, 1920, on page 2.

49. Life in Montana during this time was dominated by the Anaconda Copper Mining Company, and so it is highly likely that this was an Anaconda mine.

50. Urey, "Autobiography, 1970," 3, HCU.

51. Urey, 3. Many sources (including the finding aid to the Urey papers, Kevles, and Urey himself at times) state that Urey attended Montana State University. However, Urey's papers show that he attended the University of Montana. The confusion may arise from the fact that at this time, the university was known as the State University of Montana.

52. Brickwedde, "Harold Urey and the Discovery of Deuterium," 34.

53. Urey, "Autobiography, 1970," 4, HCU.

54. Urey, 4.

55. See Mack, "Strategies and Compromises"; and D. J. Warner, "Women Astronomers"; Rubin, *Bright Galaxies, Dark Matters*.

56. This fits with the "antielitist sentiment" prevalent in the land-grant colleges, as described in Rossiter, *Women Scientists in America*, 67.

57. Urey, "Autobiography, 1970," 4, HCU.

58. State University of Montana (Missoula), *Twenty-First Annual Register*, 26.

59. "Guide to the Morton J. Elrod Papers, 1885–1959," 93, MJE.

60. Thomas, "Harold C. Urey," 221.

61. Thomas, 222.

62. Archibald W. Bray, "Who's Who in Rensselaer Polytechnic Institute," 1936, 1, RPI.

63. *Rensselaer Polytechnic Institute Alumni News*, "Beloved Teacher," February 1943, 19, RPI.

64. *Rensselaer Polytechnic Institute Alumni News*, 19; Lowell, *Reports of the President and the Treasurer of Harvard College*, 235.

65. Alpha Delta Alpha, "The State University of Montana" 1918, p. 2, box 1, folder 1, ADA.

66. Bray, "Who's Who," 4, RPI.

67. Alpha Delta Alpha eventually became the Delta Omicron chapter of the Kappa Sigma Fraternity. "Guide to the Kappa Sigma Fraternity. Delta Omicron Chapter (State University of Montana) Records 1916–1978," n.d., ADA.

68. Everett G. Poindexter to Harold C. Urey, November 8, 1961, box 100, folder 10, HCU.

69. Urey, "Autobiography, 1970," 4, HCU.

70. University of Montana Junior Class, *The 1917 Sentinel*, 23.

71. There is a mention in one of Urey's yearbooks of the university educating "its first Chinaman." This proves that the university was not entirely white. However, the fact that this was newsworthy also reflects the rarity of nonwhite students in the student body.

72. An essay from the 1915 U of M *Sentinel* titled "Campus Lore" was accompanied by a visual comparison of the region's native past and its white present. The essay described the once thriving Flathead tribe that had called the basin their home, and the three hundred Indian lodges set up for trading when white settlers arrived. "[The tribe] had even welcomed the whites hospitably and had befriended them," the essay claimed; they had protected the settlers from more "savage" tribes. "But the Indians have been crowded out, finally. The white men needed more room." The essay concluded by celebrating the progress that white civilization had brought to Montana. University of Montana Junior Class, *The 1915 Sentinel*.

73. Jacobson, *Whiteness of a Different Color*, 41.

74. Jacobson, 42.

75. Jacobson, 42.

76. Davenport, *Heredity in Relation to Eugenics*, 202.

77. Davenport, 202.

78. Davenport, 202–3.

79. Black, *War against the Weak*, 53.

80. Jacobson, *Whiteness of a Different Color*; Paul, *Controlling Human Heredity*.

81. According to Steve Garner, "A person racialized as white can be ideologically exiled from this privilege, or may pursue values seen as antagonistic, or adhere to a minority religion, or be from another country." Garner, *Whiteness*, 11.

82. Cullen, "My Life," MCU.

83. Urey, "Autobiography, 1970," 4, HCU.

84. In DC, Bray worked under the Harvard chemist and chief of the CWS Defense Section Arthur Lamb, on biological methods of detecting gas weapons. Bray, "Who's Who," 2, RPI.

85. Shelton, "Harold Urey, Adventurer," 354.

86. Neil D. McKain, "Alpha Delta Alpha and the S.A.T.C.," 1919, p. 1, box 1, folder 1, ADA.

87. Malone, Roeder, and Lang, *Montana*, 268.

88. Van Nuys, *Americanizing the West*, 12.

89. Urey's time practicing with the dictionary is also described in Shelton, "Harold Urey, Adventurer," 351.

90. Van Nuys, *Americanizing the West*, 14.

91. Luebke, *Bonds of Loyalty*, 244.

92. Luebke, 232.

93. Luebke, 78.

94. Luebke, 226.

95. Quoted in Van Nuys, *Americanizing the West*, 51.

96. Malone, Roeder, and Lang, *Montana*, 270.

97. R. D. Bowman, *The Church of the Brethren and War, 1708–1941*, 164.

98. R. D. Bowman, 165–66.

99. Luebke, *Bonds of Loyalty*; Shenk, *"Work or Fight!,"* 60.

100. "Why Our Boys Should Be Patriotic," by Samuel Urey, is a handwritten essay in the box of family materials held by Martha (Urey) Cullen's great-granddaughter (MCU). This may have been a sermon.

101. Urey, "Autobiography, 1970," 5, HCU.

102. Warner, interview by Heitmann, 16.

103. *Science*, "Scientists in the News," 23.

104. "Alpha Delta Alpha in 1917–1918," 1918, p. 1, box 1, folder 1, ADA.

105. Urey, "Autobiography, 1970," 4–5, HCU. According to Warner, Urey was responsible for some of the more mundane aspects of work at Barrett; this is not surprising, given Urey's very junior status at the time. Warner, interview by Heitmann, 18.

106. *Chemical Engineer*, "The American Chemist Must Enlist," 41.

107. *Chemical Engineer*, 41.

108. *Chemical Engineer*, 41. The census of chemists is described in *Journal of Industrial and Engineering Chemistry*, "Chemical Warfare Service," 683.

109. *Journal of Industrial and Engineering Chemistry*, "The Parting of the Ways," 254–55.

110. Warner, interview by Heitmann, 17.

CHAPTER TWO

1. Morton John Elrod to F. C. Scheuch, n.d., box 100, folder 10, HCU.

2. Talbot, "Chemistry at the Front," 265.

3. Harrington, "American Progress in Chemical Arts," 37.

4. Elrod to Scheuch, n.d.; Harold C. Urey, "Autobiography, 1970" (unpublished manuscript), p. 5, box 1, folder 5, HCU. This change in employment was also reported in *Journal of Industrial and Engineering Chemistry*, "Personal Notes," 93.

5. Urey, interview by Heilbron, session 1, pp. 2–3.

6. With Alfred O. Nier, Baxter would in fact become interested in the question of geologic time and the application of mass spectrometry to geochemistry at least a decade earlier than Urey. Marble, "In Memoriam, Gregory Paul Baxter, 1876–1953."

7. The Carnegie Institution grant was Baxter's and was in the amount of $1,000 for the purpose of "research upon atomic weights." In addition to Jesse's work on chromium, Baxter's team also determined new weights for cadmium, manganese, bromine, lead, arsenic, iodine, silver, and phosphorus. Baxter and Jesse, "A Re-

vision of the Atomic Weight of Chromium"; Carnegie Institution of Washington, *Year Book No. 7, 1908*, 190; Scott, *Alumni Record of the University of Illinois*, 895.

8. Coffey, *Cathedrals of Science*, 54.

9. See T. W. Richards and Jesse, "The Heats of Combustion of the Octanes and Xylenes." The paper was in fact a part of Jesse's doctoral dissertation.

10. Urey, interview by Zuckerman, 9.

11. Thomas, "Harold C. Urey," 224.

12. Although the sources do not give the dates for Urey's mother's move, it seems likely that she and her husband followed Martha. C. M. Reinoehl and Eckhart, *History of an Eckhar(d)t Family*, 49–50.

13. Hildebrand, "Gilbert Newton Lewis," 494.

14. Forman, "Weimar Culture, Causality, and Quantum Theory, 1918–1927"; Nye, *Before Big Science*, chap. 7, "Nationalism, Internationalism, and the Creation of Nuclear Science, 1914–1940."

15. Heilbron and Seidel, *Lawrence and His Laboratory*, 10.

16. Heilbron and Seidel, 10.

17. Heilbron and Seidel, 8.

18. Lewis's letter to the university's president, Benjamin Ide Wheeler, in which he presents his demands is reproduced in Jolly, *From Retorts to Lasers*, chap. 10, "Lewis's First Years at Berkeley." The highest salary in the physics department belonged to F. Slate, as noted in Birge, *History of the Physics Department, Vol. I*, chap. 6, p. 6.

19. The biologist Jacques Loeb, whom Wheeler had brought to Berkeley in 1902, had by 1910 become "disillusioned with the dream of building a 'Woods Hole of the West'" and had left for the Rockefeller Institute. Servos, *Physical Chemistry from Ostwald to Pauling*, 245.

20. Coffey, *Cathedrals of Science*, xv.

21. Calvin, *Gilbert Newton Lewis*, 3.

22. Birge, *History of the Physics Department, Vol. I*, chap. 6, p. 5.

23. Calvin, *Gilbert Newton Lewis*, 3. The one exception to this trend was C. Walter Porter, who was a Berkeley graduate but had been advised by Henry C. Biddle, a faculty member hired before Lewis's arrival.

24. Hildebrand, "The Air That Urey Breathed," viii.

25. Richards quoted in Coffey, "Chemical Free Energies and the Third Law," 383.

26. Calvin, *Gilbert Newton Lewis*, 18.

27. Hildebrand, "Gilbert Newton Lewis," 493.

28. Urey, "Autobiography, 1970," 6, HCU.

29. Urey, 6. Birge stated that the graduate enrollment in chemistry during this time was in fact sixty-one students. Birge, *History of the Physics Department, Vol. II*, chap. 7, p. 34.

30. Hildebrand, “Gilbert Newton Lewis,” 492.

31. Urey, “Autobiography, 1970,” 6, HCU.

32. Hildebrand, “The Air That Urey Breathed,” ix. The German translates to “But this is *my* territory.”

33. Hildebrand, ix.

34. Jolly, *From Retorts to Lasers*, 62.

35. Calvin, *Gilbert Newton Lewis*, 7.

36. Gerald E. K. Branch quoted in Calvin, 23.

37. Seidel, “Physics Research in California,” 23.

38. See Randall, “Gilman Hall.”

39. Randall, 634–37.

40. Urey, interview by Zuckerman, 6.

41. Urey, “Statistical Distribution of the Electrons.”

42. Stanley Miller, “Harold Urey—Biographical Memoirs (Period 1923–1939)” (typescript, n.d.), p. 2, box 191, folder 9, SM.

43. Urey, “Autobiography, 1970,” 7, HCU.

44. See Urey, “Heat Capacities and Entropies”; and Urey, “Distribution of Electrons.”

45. Silvan Schweber has argued that quantum chemistry became a “quintessentially American discipline.” Schweber, “The Young John Clarke Slater and the Development of Quantum Chemistry,” 341.

46. Nye, *From Chemical Philosophy to Theoretical Chemistry*, 227.

47. This list included Robert S. Mulliken, John C. Slater, John Hasbrouck Van Vleck, Linus Pauling, Edward U. Condon, J. Robert Oppenheimer, Ralph Kronig, I. I. Rabi, Clarence Zener, David Dennison, Philip M. Morse, Henry Eyring, John G. Kirkwood, George E. Kimball, E. Bright Wilson, Hubert M. James, Francis O. Rice, and Harold C. Urey.

48. Birge, *History of the Physics Department, Vol. II*, chap. 7, page 12.

49. “Physics 220A, Quantum Theory, Fall 1921 (Williams),” and “Physics 220B, Quantum Theory, Spring 1922 (William H. Williams),” “Physics Courses” (folder), RTB. By some accounts, even after Oppenheimer arrived and began organizing a cadre of acolytes in the late 1920s, graduate students preferred Williams’s course to Oppenheimer’s. Coben, “The Scientific Establishment and Quantum Mechanics,” 459.

50. Urey, interview by Heilbron, session 1, p. 1.

51. This is a passage from Birge’s letter to his mother, dated June 26, 1922, and is included in a letter from Raymond T. Birge to Harold C. Urey, February 16, 1965, box 14, folder 1, HCU.

52. Birge enclosed this letter of recommendation in Birge to Urey, February 16, 1965.

53. See Tolman and Badger, “The Entropy of Diatomic Gases and Rotational

Specific Heat," 227. The work to which Tolman referred was Urey, "Heat Capacities and Entropies."

54. Birge to Urey, February 16, 1965.

55. Birge to Urey, February 16, 1965.

56. Nye, *From Chemical Philosophy to Theoretical Chemistry*, 228.

57. Harold C. Urey to Raymond T. Birge, February 25, 1965, box 14, folder 1, HCU.

58. Urey, interview by Heilbron, session 1, p. 7.

59. Urey, "Autobiography, 1970," 9, HCU.

60. Harold C. Urey to Gilbert N. Lewis, September 9, 1923, box 4, "Urey, Harold C. (1923–24)," GNL.

61. Urey to Lewis, September 9, 1923.

62. Harold C. Urey to Raymond T. Birge, September 2, 1923, box 28, "Urey, Harold C.," RTB.

63. Urey reported to Lewis that he finally talked to Bohr for the first time "for any length of time" at a Christmas Eve party at Bohr's home in 1923. Harold C. Urey to Gilbert N. Lewis, January 1, 1924, box 4, "Urey, Harold C. (1923–24)," GNL.

64. Urey to Lewis, January 1, 1924.

65. Davies, "American Physicists Abroad," 28. The "spirit" of the institute is similarly described in Aaserud, *Redirecting Science*, 6–15.

66. Davies, "American Physicists Abroad," 16.

67. Davies (14) characterizes Bohr's institute as a "wholly 'modernist' enclave."

68. A description of these early years of the institute and the burgeoning international community that collected there is found in Robertson, *The Early Years*, chap. 2, "A Fruitful Mysticism (1921–22)."

69. Heisenberg quoted in Davies, "American Physicists Abroad," 38.

70. Thomas, "Harold C. Urey," 225–26.

71. Davies, "American Physicists Abroad," 32.

72. Urey to Lewis, January 1, 1924.

73. Urey to Lewis, January 1, 1924.

74. Urey to Lewis, January 1, 1924.

75. Harold C. Urey to Raymond T. Birge, December 7, 1923, box 28, "Urey, Harold C.," RTB.

76. See Urey, "On the Effect of Perturbing Electric Fields."

77. Urey, "Autobiography, 1970," 9, HCU.

78. Urey, 8.

79. For an account of Friis's international activities during the early 1920s, see Erdmann, *Toward a Global Community of Historians*, chap. 7, "Overcoming Nationalism in the Study of History: Brussels, 1923."

80. Urey, "Autobiography, 1970," 8, HCU.

81. Urey to Birge, September 2, 1923.

82. Bernstein, "I. I. Rabi—I," 86.

83. Quoted in Bernstein, 88.

84. Harold C. Urey to Gilbert N. Lewis, January 1, 1924, box 4, "Urey, Harold C. (1923–24)," GNL.

85. Urey to Lewis, January 1, 1924.

86. Urey to Lewis, January 1, 1924.

87. Thomas, "Harold C. Urey," 226.

88. Quoted in Davies, "American Physicists Abroad," 59.

89. Quoted in Davies, 59.

CHAPTER THREE

1. Urey, interview by Groueff.

2. Gilbert N. Lewis to Harold C. Urey, June 9, 1926, box 4, "Urey, Harold C. (1923–24)," GNL.

3. Harold C. Urey to Gilbert N. Lewis, April 8, 1924, box 4, "Urey, Harold C. (1923–24)," GNL.

4. Harold C. Urey to Gilbert N. Lewis, May 4, 1924, box 4, "Urey, Harold C. (1923–24)," GNL.

5. Harold C. Urey to Gilbert N. Lewis, July 3, 1924, box 4, "Urey, Harold C. (1923–24)," GNL.

6. Harold C. Urey to Edwin C. Kemble, June 2, 1924, box 49, folder 4, HCU.

7. Urey thanked Lewis for "securing this position" in a letter sent from Baltimore September 30, 1924, box 4, "Urey, Harold C. (1923–24)," GNL.

8. Harold C. Urey to Gilbert N. Lewis, July 3, 1924, box 4, "Urey, Harold C. (1923–24)," GNL.

9. Harold C. Urey to Gilbert N. Lewis, September 30, 1924, box 4, "Urey, Harold C. (1923–24)," GNL.

10. Harold C. Urey to Edwin C. Kemble, June 18, 1925, box 49, folder 4, HCU.

11. Harold C. Urey to Edwin C. Kemble, February 18, 1926, box 49, folder 4, HCU.

12. Arnold, Bigeleisen, and Hutchison, "Harold Clayton Urey," 390–91.

13. Urey, "The Teaching of Atomic Structure," 284.

14. Urey, "Discussion," 258.

15. Urey, "The Teaching of Atomic Structure," 284.

16. Urey, 284.

17. Urey, 284.

18. Urey, 284.

19. Urey, 284–85.

20. Urey, interview by Heilbron, session 1, p. 18. Ironically, Ames would later go on to play a role in the founding of NASA.

21. Urey, interview by Heilbron, session 1, pp. 15–16.

22. Harold C. Urey, "Autobiography, 1970" (unpublished manuscript), p. 10, box 1, folder 5, HCU.

23. Smallwood, review of *Atoms, Molecules and Quanta*, 2588.

24. Gilbert N. Lewis to Harold C. Urey, August 27, 1926, box 4, "Urey, Harold C. (1923–24)," GNL.

25. Harold C. Urey to Raymond T. Birge, January 12, 1929, RTB.

26. Nye, *From Chemical Philosophy to Theoretical Chemistry*, 252.

27. Kevles, *The Physicists*, 225.

28. Bernstein, "I. I. Rabi—I," 94.

29. Bernstein, 94.

30. Urey, interview by Heilbron, session 1, p. 13.

31. See Harold C. Urey, "Structure of the Hydrogen Molecule Ion."

32. Urey's original work on the Zeeman effect was published as "On the Effect of Perturbing Electric Fields on the Zeeman Effect of the Hydrogen Spectrum." His work with Bichowsky resulted in their 1926 paper, "A Possible Explanation of the Relativity Doublets and Anomalous Zeeman Effects by Means of a Magnetic Electron."

33. See Uhlenbeck and Goudsmit, "Ersetzung Der Hypothese Vom Unmechanischen Zwang."

34. Urey and Bichowsky, "Possible Explanation of Relativity Doublets," 80.

35. Urey, interview by Heilbron, session 1, p. 24.

36. See Washburne, "Black Body Radiation and Decomposition of Nitrogen Pentoxide"; and Dawsey, "Absorption Spectrum and Photochemical Dissociation."

37. See Lavin, "Reactions of Atomic Hydrogen"; and Smallwood, "Study of Active Modifications of Hydrogen."

38. Harold C. Urey to Raymond T. Birge, December 28, 1928, RTB.

39. Urey, interview by Heilbron, session 1, p. 11.

40. Urey, interview by Heilbron, session 1, p. 14.

41. Harold C. Urey to Gilbert N. Lewis, December 17, 1931, box 4, "Urey, Harold C. (1923–24)," GNL.

42. Harold C. Urey to Raymond T. Birge, February 6, 1929, RTB.

43. Urey, "Autobiography, 1970," 10, HCU.

44. Stanley Miller, "Harold Urey—Biographical Memoirs (Period 1923–1939)" (typescript, n.d.), p. 5, box 191, folder 9, SM.

45. See Murphy and Urey, "On the Relative Abundances of the Nitrogen and Oxygen Isotopes," 141; and Urey and Murphy, "The Relative Abundance of N14 and N15," 575.

46. See Urey and Johnston, "Absorption Spectrum of Chlorine Dioxide," 2131.

47. See Urey and Bradley, "Raman Spectra of Silico-Chloroform," 843.

48. Pitzer and Shirley, "William Francis Giauque, 1895–1982," 43.

49. See Giauque and Johnston, "An Isotope of Oxygen, Mass 18"; Giauque and

Johnston, "The Heat Capacity of Oxygen"; and Giauque and Johnston, "An Isotope of Oxygen, Mass 17."

50. Aston, "Bakerian Lecture," 487.

51. Urey, "Autobiography, 1970," 11, HCU.

52. Urey's method required that hydrogen be distilled at its triple point—the temperature and pressure at which the solid, liquid, and gas phases of hydrogen can coexist in thermodynamic equilibrium.

53. Urey, interview by Heilbron, session 2.

54. Brickwedde, "Atomic Models as Proposed by Bohr."

55. Urey later learned that this commercially prepared hydrogen had in fact been enriched. It had been prepared through the electrolysis of water, which favors molecular H-2. In 1933 Lewis prepared the first highly concentrated samples of heavy water using electrolysis. Lewis and MacDonald, "Concentration of H2 Isotope."

56. Miller, "Harold Urey—Biographical Memoirs (Period 1923–1939)," 10, SM.

57. Bigeleisen quoted in Coffey, *Cathedrals of Science*, 220.

58. Urey, "Autobiography, 1970," 12–13, HCU.

59. Harold C. Urey to Raymond T. Birge, June 21, 1968, box 14, folder 1, HCU.

60. Harold C. Urey to Gilbert N. Lewis, April 11, 1933, box 4, "Urey, Harold C. (1923–24)," GNL.

61. Harold C. Urey to Raymond T. Birge, May 21, 1934, box 28, Letters Written to Birge, RTB.

62. Coffey, *Cathedrals of Science*, 221. Coffey also speculates that Lewis's difficult personality may have kept him from winning the Nobel Prize despite having been nominated several times. According to Coffey, Walther Nernst and Lewis shared a lifelong aversion to one another; Nernst's associates on the Nobel Committee blocked Lewis's selection. Ibid., 191–207.

63. Harold C. Urey to Kenneth S. Pitzer, April 3, 1967, box 74, folder 29, HCU.

64. Harold C. Urey to Kenneth S. Pitzer, March 27, 1967, box 74, folder 29, HCU.

65. Urey, interview by Heilbron, session 2, p. 3. Family reasons may have once again played a part in Urey's desire to move west. The majority of Urey's family, including his aging mother, were now living in California. As the next chapter reveals, when given the opportunity to explore possible locations for a new Institute for Nuclear Studies, Urey first explored universities on the West Coast.

66. Kuznick, *Beyond the Laboratory*, 101.

67. Cohn, interview by Gortler, 35.

68. Cohn, interview by Gortler, 33.

69. The physicist Jerrold R. Zacharias recalled that Columbia University administrators "were willing to have [Jews as] graduate students, but not to have

[Jews as] faculty. And this was true of Bell Labs and all other universities." Quoted in Rigden, *Rabi*, 104.

70. Bernstein, "I. I. Rabi—I," 86.
71. Cohn, interview by Gortler, 35.
72. Cohn, interview by Gortler, 38.
73. Cohn, interview by Gortler, 38.
74. Cohn, interview by Gortler, 29.
75. Cohn, interview by Gortler, 33.
76. Cohn, "Harold Urey: A Personal Remembrance," 9.
77. Coffey, *Cathedrals of Science*, 219.
78. Bernstein, "I. I. Rabi—I," 96.
79. Bernstein, 50.

CHAPTER FOUR

1. *New York Times*, "Bid Farewell to Urey."
2. M. H. Hall, "All Sides of Harold Urey," 54.
3. Stanley Miller, "Harold Urey—Biographical Memoirs (Period 1923–1939)" (typescript, n.d.), pp. 13–14, box 191, folder 9, SM.
4. See Kohler, "Rudolf Schoenheimer, Isotopic Tracers, and Biochemistry"; and Creager, *Life Atomic*. Creager's work suggests that, had radioactive isotopes not overtaken stable isotopes as tracers in biochemical research, Urey's contributions to biology would figure as an even more important part of his legacy.
5. The chemistry profession's efforts to promote their usefulness to industry and society are chronicled in Rhees, "The Chemists' Crusade"; the profession's similar efforts in wartime, and the war's impact on their activities, are examined in Rhees, "The Chemists' War."
6. The efforts of the editor of the *Journal of Industrial and Engineering Chemistry* during this time are described in Reed, *Crusading for Chemistry*, chap. 5, "The Mouthpiece of Chemistry."
7. An account of the attacks against science during the 1930s can be found in Kevles, *The Physicists*, chap. 16, "Revolt against Science." Accounts of the defense of science by scientists during this time can be found in Kuznick, *Beyond the Laboratory*, 9–37.
8. Kevles, *The Physicists*, 239.
9. Whittemore, "World War I, Poison Gas Research," 139.
10. Whittemore, 140.
11. Whittemore, 145. The language used to describe the "public man" reflected and reinforced existing gender prejudices.
12. Urey, "Accomplishments and Future of Chemical Physics," 226.
13. Urey, 223.

14. Urey, 226.

15. Harold C. Urey, "Autobiography, 1970" (unpublished manuscript), p. 40, box 1, folder 5, HCU.

16. Urey, "Chemistry and the Future," 139.

17. Harold C. Urey, "Science and the Humanities" June 3, 1935, p. 14, box 140, folder 7, HCU.

18. Urey, "Science and the Humanities," 14, HCU.

19. Urey, "Chemistry and the Future," 135.

20. Urey, 135.

21. Urey, 139. Although he listed Russia's turn toward communism as one of the "serious difficulties" of the century, Urey (not unlike other observers of the Depression) looked favorably on the Bolsheviks' stated commitment to distributing the fruits of mass production to all citizens of the Soviet Union.

22. Urey, 139.

23. Urey, 135.

24. Urey, 134.

25. Urey, "Accomplishments and Future of Chemical Physics," 225.

26. Urey, "Chemistry and the Future," 134.

27. Urey, 134.

28. Urey, "Accomplishments and Future of Chemical Physics," 225.

29. Urey, 223.

30. On the history of the argument that science and liberal government are linked, see Aronova, "Studies of Science before 'Science Studies'"; Aronova and Turchetti, *Science Studies during the Cold War and Beyond*; and Aronova, "The Congress for Cultural Freedom."

31. Hollinger, "The Defense of Democracy."

32. Hollinger, *Science, Jews, and Secular Culture*, 82.

33. Urey, "Accomplishments and Future of Chemical Physics," 223.

34. Urey, 227.

35. Urey, 227, 224.

36. Urey, "Science and the Humanities," 2, HCU.

37. Urey, 2.

38. Urey, "Chemistry and the Future," 134.

39. Urey, "Science and the Humanities," 10–11, HCU.

40. Urey, "Chemistry and the Future," 134.

41. Bryson and Finkelstein, *Science, Philosophy, and Religion*, 211.

42. Urey participated in the founding of the conference and remained on the board for several years as it became an annual event, even as his participation dwindled. For the history of the conference and a more complete analysis of its aims, see Buettler, "Organizing an American Conscience"; and Cohen-Cole, *The Open Mind*, 13–34.

43. Descriptions of other such manifestations can be found in Hollinger, "The Defense of Democracy"; Hollinger, "Science as a Weapon."

44. Louis Finkelstein, memorandum, September 12, 1940, p. 3, record group 5, box 1, folder 5A-1-5, CSPR.

45. Van Wyck Brooks, "Statement Oct 1940 V. W. Brooks," October 1940, p. 1, record group 5, box 1, folder 5A-1-12, CSPR.

46. Brooks, 2.

47. "Executive Committee—Press Release Draft," August 9, 1940, p. 1, record group 5, box 1, folder 5A-1-5, CSPR.

48. "Executive Committee—Press Release Draft," 2.

49. "Executive Committee—Press Release Draft," 2.

50. "Executive Committee—Press Release Draft," 3.

51. Louis Finkelstein to Harold C. Urey, March 29, 1940, 1A-27-16, JTSA; Harold C. Urey to Louis Finkelstein, April 1, 1940, 1A-27-16, JTSA; *New York Times*, "Conference to Aid Democratic Ideals."

52. Louis Finkelstein to Harold C. Urey, September 18, 1940, 1A-27-16, JTSA.

53. "Minutes of the Board of Directors of the Conference on Science, Philosophy and Religion in Their Relation to the Democratic Way of Life, Inc.," January 13, 1941, p. 6, record group 5, box 1, folder 5A-1-2, CSPR; Bryson and Finkelstein, *Science, Philosophy, and Religion*, 551–52.

54. "Monday AM + AFT Sessions," September 8, 1941, pp. 37–38, box 36, folder "2nd Conference," CSPR.

55. "Monday AM + AFT Sessions," 38.

56. "Monday AM + AFT Sessions," 19–20.

57. See Buettler, "Organizing an American Conscience."

58. "Monday AM + AFT Sessions," 111.

59. "Monday AM + AFT Sessions," 113–14.

60. Schoen, "The Basis for Faith in Democracy," 98.

61. "Monday AM + AFT Sessions," 101–2.

62. *New York Times*, "Berle Sees Nazis Losing Strength."

63. Meitner and Frisch, "Disintegration of Uranium by Neutrons," 239.

64. An account of physicists' increasing anxiety and their calls for action is found in Rhodes, *The Making of the Atomic Bomb*, chap. 10, "Neutrons." Rhodes's book is in general one of the most complete histories of the American atomic bomb program.

65. Hewlett and Anderson, *The New World, 1939–1946*, 24.

66. Smyth, "Atomic Energy for Military Purposes," 375.

67. The most extensive account of Urey's wartime activities and attitudes was written by his wartime research assistant, Karl P. Cohen. See Cohen, "Harold C. Urey, the War Years: 1939–1944" (typescript, March 21, 1983), p. 1, box 191, folder 9, SM. Unfortunately, Cohen's account was published only in abbreviated

form as Cohen, "Harold C. Urey 1893–1981"; see also Cohen et al., "Harold Clayton Urey."

68. Although no correspondence between Urey and Aage Friis survives from this period, it is worth noting that Friis was involved in these same activities in Denmark. Bohr and Aaserud, *Popularization and People (1911–1962)*, 435.

69. Fermi, *Atoms in the Family*, 145.

70. Harold C. Urey to George de Hevesy, March 21, 1940, box 42, folder 27, HCU.

71. Harold C. Urey to Warren Weaver, May 30, 1940, box 42, folder 27, HCU; Warren Weaver to Harold C. Urey, June 7, 1940, box 42, folder 27, HCU.

72. Harold C. Urey to Linus Pauling, June 16, 1938, box 419.1, folder 1, LP. In this letter to Pauling, Urey also indicated that he was working with the Institute of International Education to help secure grants for the fleeing scientists, but that the official grant proposals had to come from sponsoring institutions.

73. Cohen, "Harold C. Urey, the War Years," 3, SM.

74. Urey, interview by Groueff.

75. Cohen, "Harold C. Urey, the War Years," 5, SM.

76. Thomas, "Harold C. Urey," 233.

77. Cohen, "Harold C. Urey, the War Years," 10, SM.

78. Arnold, "Harold C. Urey," 3, SM.

79. Urey, interview by Heilbron, session 2, p. 3.

80. Brian, *The Voice of Genius*, 260.

81. Groves, interview by Groueff, January 5, 1965.

82. Urey, interview by Groueff.

83. Groves, interview by Groueff, January 5, 1965.

84. Nichols, interview by Groueff, January 4, 1965.

85. Cohen, "Harold C. Urey, the War Years," 6, SM.

86. Cohen, 3.

87. Cohen, 4.

88. Smyth, "Atomic Energy for Military Purposes," 376.

89. Cohen, "Harold C. Urey, the War Years," 5, SM.

90. Quoted in Hershberg, *James B. Conant*, 489.

91. Cohen, "Harold C. Urey, the War Years," 9–10, SM.

92. Cohen, 12.

93. Taylor, interview by Groueff.

94. Groves, interview by Groueff, January 5, 1965.

95. Fermi, *Atoms in the Family*, 202.

96. Groves, interview by Groueff, January 5, 1965; Taylor, interview by Groueff.

97. Nichols, interview by Groueff, January 4, 1965.

98. Nichols, interview by Groueff, January 4, 1965.

99. Silverstein, *Harold Urey*, 51.

100. Urey, interview by Groueff.

101. Urey, interview by Zuckerman, 19.

102. Urey, interview by Zuckerman, 19.

CHAPTER FIVE

1. Teller was then faculty at George Washington University, but had moved to Urey's Columbia University SAM Lab in 1941 before becoming attached to Compton's University of Chicago Metallurgical Laboratory the following year.

2. Enrico Fermi to Harold C. Urey, July 11, 1945, box 34, folder 4, HCU.

3. Harold C. Urey to Edward Teller, July 7, 1945, box 90, folder 2, HCU.

4. Harold C. Urey to Enrico Fermi, July 5, 1945, box 34, folder 4, HCU.

5. Urey to Teller, July 7, 1945.

6. In a volume dedicated to Urey on his 75th birthday, Hutchins reminisced, "It seems to me, as I look back on it, that I spent most of my time at the University of Chicago trying to persuade Harold Urey to join the faculty." Hutchins, "The Man We Love," v.

7. McGrayne, *Nobel Prize Women in Science*, 175–200.

8. Urey to Teller, July 7, 1945.

9. Urey to Teller, July 7, 1945.

10. Samuel K. Allison, "Thoughts on 10th Anniversary of First Chain Reaction" (typescript, November 7, 1952), p. 8, box 25, folder 7, SKA.

11. Fermi, *Atoms in the Family*, 247.

12. Dunning went on to serve as dean of engineering from 1950 to 1969. Urey's view of Dunning as more of a political operator and self-promoter than a competent scientist was vindicated by Dunning's time as dean, which coincided with a decline in the School of Engineering's reputation and national standing, even as he managed to secure the support of donors and used this support to prevent his forced resignation. McCaughey, *Stand, Columbia*, 354.

13. Robert S. Mulliken to Harold C. Urey, August 3, 1945, box 56, folder 9, HCU.

14. Harold C. Urey to Joseph E. Mayer and Maria Goeppert Mayer, August 18, 1945, box 56, folder 9, HCU.

15. Harold C. Urey to Walter Bartky, August 6, 1945, box 12, folder 29, HCU.

16. Samuel K. Allison to Don M. Yost, December 4, 1945, box 3, folder 3, HCU.

17. Allison, "Institute for Nuclear Studies, the University of Chicago," 482.

18. Allison, 484.

19. Samuel K. Allison, "Research Corp. Dinner" (typescript, March 18, 1952), box 25, folder 13, SKA.

20. Samuel K. Allison, "Description of INS Activities" (typescript, December 9, 1953), box 24, folder 8, SKA.

21. The word *traumatized* is no overstatement—it is precisely the word Urey's colleagues chose to use in describing his postwar state of mind. See Cohen et al., "Harold Clayton Urey," 643.

22. Hutchins, "The Man We Love," v.

23. Joseph Mayer, "[Biography of Harold C. Urey]" (typescript, 1970), box 1, folder 11, HCU 9.

24. Hans E. Suess, "Harold C. Urey" (copy, December 1981), p. 1, box 10, folder 12, SE.

25. James R. Arnold, "Harold C. Urey Chair in Chemistry, Inaugural Address" (typescript, April 29, 1983), box 139, folder 9, SM.

26. Quoted in *New Yorker*, "The Talk of the Town," 23.

27. Quoted in *New Yorker*, 24.

28. Lang, "That's Four Times 10^{-4} Ergs," 84.

29. The scientists' movement after the war is described in Smith, *A Peril and a Hope*; and Boyer, *By the Bomb's Early Light*.

30. *New Yorker*, "The Talk of the Town," 23.

31. Quoted in *New Yorker*, 23.

32. Quoted in *New Yorker*, 24.

33. Urey and Amrine, "I'm a Frightened Man," 18.

34. Urey and Amrine, 51.

35. Urey and Amrine, 51.

36. "Proceedings: The Social Task of the Scientist in the Atomic Era—A Symposium, Emergency Committee of Atomic Scientists," November 17, 1946, p. 12, box 3.015, folder 15.4, LP.

37. The tendency toward world government among atomic scientists during this period is described in Boyer, *By the Bomb's Early Light*, chap. 3, "Atomic-Bomb Nightmares and World-Government Dreams."

38. Urey, "Science and the Humanities," 14.

39. Baratta, *The Politics of World Federation*, chap. 15, "Robert M. Hutchins: Framing a World Constitution."

40. See Committee to Frame a World Constitution, *Preliminary Draft of a World Constitution*.

41. Urey and Amrine, "I'm a Frightened Man," 51.

42. Urey, "A Scientist Views the World Situation," 4.

43. Urey, 4.

44. Urey, 4.

45. Acheson and Lilienthal, *International Control of Atomic Energy*.

46. Harold C. Urey to Atomic Scientists of Chicago, September 30, 1946, box 11, folder 5, HCU.

47. Urey to Atomic Scientists of Chicago, September 30, 1946.

48. The split between the two men was obvious by the end of the 1940s, but had begun developing earlier. Urey and Szilard's difference of opinion and their leadership roles in defining those positions for their scientific colleagues is described in Hawkins, Greb, and Szilard, *Toward a Livable World*, 4. Streit had been proposing an Atlantic Union in order to preserve peace in the face of fascism since as early as 1939; after World War II, the plan was adapted to the problems of the Cold War. Streit, *Union Now*; Streit, *Union Now with Britain*.

49. Quoted in Hawkins, Greb, and Szilard, *Toward a Livable World*, 57.

50. Hawkins, Greb, and Szilard, 58.

51. Hawkins, Greb, and Szilard, 58.

52. Urey, "Should America Build the H-Bomb?," 73.

53. Harold C. Urey to Philip Iseley, August 9, 1957, box 47, folder 1, HCU.

54. Thomas quoted in J. Wang, *American Science in an Age of Anxiety*, 48; J. Wang, chap. 2, "Fear, Suspicion, and the Surveillance State: The FAS, HUAC, and the FBI, 1945–1948."

55. J. Wang, "Science, Security, and the Cold War," 239.

56. Gellhorn, *Security, Loyalty, and Science*, 1.

57. Harold C. Urey to Albert Einstein, April 15, 1947, box 1, folder 13, ECAS.

58. Harold C. Urey to Charles A. Kraus, March 23, 1948, box 18, folder 33, ECAS.

59. Harold C. Urey to Frank Jewett, March 23, 1948, box 18, folder 33, ECAS.

60. *New York Times*, "Brooklyn Professor, Called a Red, Bars House Queries on Atom Data," 1.

61. Quoted in Eckel, "Red Inquiry Hears Attack," 30. Rushmore also listed several other University of Chicago professors, including the philosopher Rudolf Carnap, as belonging to Communist front organizations.

62. *New York Times*, "Secrecy on Atom Called Excessive," 8; Weart, "Scientist Scores Secrecy on Atom," 5.

63. Harold C. Urey, "Telegram to Trustees," October 13, 1948, box 1, folder 16, ECAS.

64. Quoted in Cohen et al., "Harold Clayton Urey," 643.

65. C. E. Hennrich to A. H. Belmont, memorandum, June 21, 1950, Harold Urey, file 116-18315, vol. 1, Federal Bureau of Investigation, Washington, DC.

66. The "Communist front and Communist dominated organizations" to which the FBI determined Urey belonged included the American Committee for Democracy and Intellectual Freedom, the American Committee for Protection of Foreign Born, the American Committee to Save Refugees, the Coordination Committee to Lift the Embargo on Loyalist Spain, the Friends of Abraham Lincoln Brigade, the Greater New York Emergency Conference of Inalienable Rights, the League of American Writers, the National Council of American-Soviet Friendship, the National Emergency Conference for Democratic Rights, the Spanish Intellec-

tual Aid Committee, the Veterans of Abraham Lincoln Brigade, the Medical Bureau and North American Committee to Aid Spanish Democracy, and the National Federation for Constitutional Liberties.

67. Harold C. Urey to Louis Finkelstein, August 11, 1949, box 44, folder 11, HCU.

68. Harold C. Urey, "Chemically Pure," 65.

69. Urey to Finkelstein, August 11, 1949.

70. In Conant's defense, he did firmly oppose the loyalty oath and, as a result, made a personal enemy of the FBI director J. Edgar Hoover. Hershberg, *James B. Conant*, 416.

71. Hershberg, 418.

72. Harold C. Urey to Joel H. Hildebrand, August 10, 1959, box 43, folder 6, HCU.

73. Harold C. Urey to Harry S. Truman, December 16, 1952, box 90, folder 37, HCU.

74. Harold C. Urey to Edward U. Condon, January 29, 1953, box 21, folder 12, HCU.

75. Urey's telegram to Eisenhower is reproduced in US House of Representatives Committee on Un-American Activities, *Trial by Treason*, 33.

76. Report, Chicago, October 28, 1953, Harold Urey, file 116-18315, vol. 1, Federal Bureau of Investigation, Washington, DC.

77. Report, October 28, 1953.

78. Report, October 28, 1953.

79. US House of Representatives Committee on Un-American Activities, *Trial by Treason*, 33–34.

80. Cohen et al., "Harold Clayton Urey," 643.

81. Harold C. Urey, "The Intellectual Revolution—Duke University Commencement Address" (typescript, June 1, 1953), p. 1, box 140, folder 5, HCU. Here Urey likely had in mind the Nazi's distortions of anthropology, their concept of the "master race," and their fascination with eugenics.

82. Urey, 4.

83. Urey, 2.

84. Urey, 5.

85. Urey, 5.

86. Urey, 6.

87. Urey, 5.

88. Urey, 9.

89. Urey, 10.

90. Urey, 10.

91. Urey, 11.

92. Urey, 11.

93. Harold C. Urey, "Evolution vs. Miraculous Creation," n.d., p. 14, box 144, folder 25, HCU.

94. Urey, "The Intellectual Revolution," 14, HCU.

95. Urey, 14.

96. Urey, 15.

CHAPTER SIX

1. Hutchison, "Harold Clayton Urey, 1893–1981," 520.

2. Urey, "Thermodynamic Properties of Isotopic Substances," 578.

3. Urey, 581.

4. Brush, *Fruitful Encounters*, 144.

5. Samuel Epstein, untitled manuscript, January 28, 1981, p. 2, box 10, folder 12, SE.

6. Doel, *Solar System Astronomy in America*, 79.

7. White, *Mass Spectrometry in Science and Technology*, 11.

8. Nier, "Some Reminiscences of Mass Spectrometry," 387; Grayson, *Measuring Mass*, 16–17.

9. Harold C. Urey to Walter Bartky, August 6, 1945, box 12, folder 29, HCU.

10. Harold C. Urey to Leo Samuels, November 24, 1945, box 69, folder 21, HCU.

11. Harold C. Urey to A. V. Peterson, September 5, 1946, box 74, folder 8, HCU.

12. Alfred O. C. Nier to A. V. Peterson, October 1, 1946, box 69, folder 21, HCU.

13. A. V. Peterson to Harold C. Urey, October 23, 1946, box 74, folder 8, HCU.

14. Harold C. Urey to Alfred O. C. Nier, October 7, 1946, box 69, folder 21, HCU.

15. Alfred O. C. Nier to Philip Owen, July 10, 1946, box 69, folder 21, HCU.

16. Joseph Ney to Harold C. Urey, July 9, 1946, box 69, folder 21, HCU; Harold C. Urey to Joseph Ney, July 12, 1946, box 69, folder 21, HCU.

17. Alfred O. C. Nier to Committee on Growth, December 18, 1946, box 69, folder 21, HCU.

18. Alfred O. C. Nier to Harold C. Urey, February 11, 1947, box 69, folder 21, HCU.

19. Nier to Committee on Growth, December 18, 1946.

20. Nier to Committee on Growth, December 18, 1946.

21. Harold C. Urey to Alfred O. C. Nier, February 28, 1947, box 69, folder 21, HCU.

22. Samuel Epstein to Harold C. Urey, July 17, 1947, box 31, folder 2, HCU.

23. Harold C. Urey to Samuel Epstein, July 23, 1947, box 31, folder 2, HCU.

24. Harold C. Urey to Samuel Epstein, March 4, 1949, box 31, folder 2, HCU.

25. Fisher, *The Seventy Years of the Department of Geology*, 52; Goldsmith, "Some Chicago Georecollections," 4.

26. Fisher, *The Seventy Years of the Department of Geology*, 58.

27. Harrison Brown, "Formal Establishment of Training in Geochemistry," April 4, 1950, p. 4, box 15, folder 37, HCU.

28. Goldsmith, "Some Chicago Georecollections," 9.

29. Fisher, *The Seventy Years of the Department of Geology*, 61.

30. See Epstein et al., "Carbonate-Water Isotopic Temperature Scale"; Urey, Lowenstam, et al., "Measurement of Paleotemperatures."

31. Doel, Levin, and Marker, "Extending Modern Cartography to the Ocean Depths," 608; Oreskes, *Science on a Mission*.

32. Doel, "The Earth Sciences and Geophysics," 398.

33. Mullaney, "Atomic Research Shows Progress."

34. H. Gershinowitz to E. G. Gaylord, April 8, 1947, box 6, folder 1, HCU.

35. Harold C. Urey to T. V. Moore, June 25, 1947, box 86, folder 36, HCU.

36. Reinhardt, "The Chemistry of an Instrument," 231; Reinhardt, *Shifting and Rearranging*.

37. C. A. Young to Harold C. Urey, March 9, 1948, box 6, folder 1, HCU.

38. Harold C. Urey, "Penrose Application," July 17, 1947, box 37, folder 10, HCU.

39. Henry Aldrich to Harold C. Urey, September 22, 1948, box 37, folder 10, HCU.

40. American Petroleum Institute to Harold C. Urey, January 7, 1952, box 6, folder 1, HCU.

41. Rainger, "Science at the Crossroads," 366. As Rainger points out, this meant that "studies of ocean bottoms, surface layers, coastlines, and almost any topic other than biological oceanography were, at once, both intellectually meaningful to oceanographers and operationally useful to the Navy" (ibid.).

42. Harold C. Urey, "Proposal for Task Order under Contract No. N6 Ori-20," November 24, 1948, box 94, folder 19, HCU.

43. Urey, "Proposal for Task Order," HCU.

44. Office of Naval Research to Harold C. Urey, August 22, 1952, box 95, folder 1, HCU.

45. Harold C. Urey, "Proposal for Extension of Research on the Natural Abundance of Deuterium and Other Isotopes in Nature," February 1953, box 92, folder 9, HCU; National Science Foundation to Harold C. Urey, July 20, 1953, box 68, folder 1, HCU.

46. George M. Cableman to Roy B. Snapp, "Implementing Action—Dr. Harold C. Urey," n.d., Atomic Energy Commission, Records, box 8, folder 095 (12-11-46), National Archives and Records Administration, Washington, DC.

47. Creager, "Atomic Tracings."

48. Harold C. Urey to Kenneth S. Pitzer, January 1, 1951, box 92, folder 8, HCU.

49. Kenneth S. Pitzer to Harold C. Urey, February 9, 1951, box 92, folder 8, HCU.

50. See Harrison Brown, "A Proposed Program for the Accumulation of Quantitative Data Concerning: The Chemical Composition of Meteorites and the Earth's Crust; the Relative Abundances of Elements in the Solar System; the Ages of the Elements and Planets," 1949, box 15, folder 37, HCU; and Harold Urey, "Atomic Energy Commission Contract No. AT(11-1)-101 to Investigate the Natural Abundance of Deuterium and Other Isotopes in Nature," February 1953, box 92, folder 9, HCU.

51. Harold C. Urey to David Feinman, August 5, 1977, box 35, folder 24, HCU.

52. Garrison et al., "Reduction of Carbon Dioxide."

53. Urey, "On the Early Chemical History of the Earth," 351.

54. Harold C. Urey, "Report on Contract AT(11-1)-101," February 15, 1953, p. 1, box 92, folder 9, HCU.

55. Urey, "On the Early Chemical History of the Earth," 362.

56. Urey to Feinman, August 5, 1977.

57. Urey to Feinman, August 5, 1977.

58. Harold C. Urey to Edward U. Condon, January 29, 1953, box 21, folder 12, HCU.

59. Harold C. Urey to *Science Magazine*, telegram, [March 10, 1953], box 58, folder 18, HCU.

60. Howard Meyerhoff to Stanley L. Miller, March 11, 1953, box 58, folder 18, HCU.

61. Miller, "Production of Amino Acids," 528.

62. Urey to Feinman, August 5, 1977.

63. Trytten, *Student Deferment in Selective Service*, 84.

64. Flynn, "The Draft and College Deferments."

65. Harold C. Urey to Local Board no. 48 [Alameda County, California], August 28, 1953, box 58, folder 18, HCU.

66. For a treatment of Oparin's influence on the emerging exobiology community, see Dick and Strick, *The Living Universe*, chap. 2, "Organizing Exobiology."

67. Harold C. Urey to Stanley L. Miller, May 3, 1956, box 58, folder 18, HCU.

68. Stanley L. Miller to Harold C. Urey, April 10, 1957, box 58, folder 18, HCU.

69. Miller to Urey, April 10, 1957.

70. Harold C. Urey to Stanley L. Miller, April 13, 1957, box 58, folder 18, HCU.

CHAPTER SEVEN

1. Harold C. Urey to Harry G. Thode, August 4, 1975, box 90, folder 14, HCU.

2. *New York Times*, "Urey Prefers Shell Work."

3. Harold C. Urey, tape 7a, interview by Ian I. Mitroff, December 1, 1969, box 1, folder 7, IIM.

4. Harold C. Urey to Rollin D. Hemens, November 9, 1950, box 12, folder 14, HCU.

5. Harold C. Urey, "Autobiography, 1970" (unpublished manuscript), p. 24, box 1, folder 5, HCU.

6. Urey, 24.

7. Urey, *The Planets*, ix; Slichter, "Cooling of the Earth," 598.

8. Urey concluded that the abundance of radioactive elements within the primitive Earth were sufficient to heat the planet, based on new radioactive abundance measurements collected in Brown's meteorite studies at the institute. Urey's paper, "A Hypothesis Regarding the Origin of the Movements of the Earth's Crust," was presented at the 1949 meeting of the NAS: National Academy of Sciences, "Abstracts of Papers Presented at the 1949 Autumn Meeting." The contents of Urey's presentation were also reported in Plumb, "Earth Heating Up, Urey Theory Holds."

9. According to Doel, this conference was "hastily arranged in January 1950 in part to provide a forum for discussing Urey's ideas." Doel, *Solar System Astronomy in America*, 95, 97.

10. Doel, 96.

11. In attendance at Rancho Santa Fe were Harrison Brown, Linus Pauling, Beno Gutenberg, David T. Griggs, Patrick M. Hurley, Francis Birch, W. F. Latimer, James Gilluly, Adolph Knopf, William W. Rubey, Edward Teller, Howard P. Robertson, and Fred Whipple. Many of the fruits of these discussions can be found in Urey, "The Origin and Development of the Earth and Other Terrestrial Planets"; and Urey, "The Origin and Development of the Earth and Other Terrestrial Planets: A Correction."

12. For biographical information on Suess, see Waenke and Arnold, "Hans E. Suess."

13. See Urey, "Chemical Fractionation in the Meteorites"; Urey, "The Abundances of the Elements"; Urey, "The Cosmic Abundances of Potassium, Uranium, and Thorium"; and Suess and Urey, "Abundances of the Elements."

14. Brush, "Theories of the Origin of the Solar System," 52. Some details of Urey and Kuiper's collaboration and falling out are presented throughout Doel, *Solar System Astronomy in America*.

15. The failed collaboration between Urey and Kuiper is described in Doel, *Solar System Astronomy in America*, chap. 4, "Consensus, Then Controversy."

16. Doel, 148.

17. Harold C. Urey to Dinsmore Alter, February 9, 1955, box 3, folder 6, HCU.

18. Urey, "Some Criticisms of 'On the Origin of the Lunar Surface Features,'" 423.

19. Harold C. Urey to Dinsmore Alter, June 9, 1954, box 3, folder 6, HCU.

20. Urey, "Some Criticisms of 'On the Origin of the Lunar Surface Features,'" 423.

21. Harold C. Urey, "The Intellectual Revolution—Duke University Commencement Address" (typescript, June 1, 1953), p. 10, box 140, folder 5, HCU.

22. Doel, *Solar System Astronomy in America*, 142.

23. These remembrances are shared on the website of the Lunar and Planetary Laboratory. See Cruikshank, Coyne, and Binder, "Gerard P. Kuiper."

24. Cruikshank, "Gerard Peter Kuiper, December 7, 1905–December 24, 1973," 278.

25. Urey, "Diamonds, Meteorites, and the Origin of the Solar System," 623.

26. Urey described these ideas developed at Oxford in Harold C. Urey, "Statement before Committee of the House of Representatives on the Objectives of the Exploration of the Moon and Planets" (typescript, March 11, 1965), HCU NASA.

27. Harold C. Urey to Willard F. Libby, August 26, 1957, box 52, folder 45, HCU.

28. Harold C. Urey to Willard F. Libby, November 22, 1957, box 52, folder 45, HCU.

29. Harold C. Urey to Willard F. Libby, November 18, 1958, box 52, folder 45, HCU.

30. *New York Times*, "Scientific Romancing," E8.

31. *New York Times*, "Scientific Romancing."

32. *New York Times*, "Scientific Romancing."

33. Not long after the *Times* discovered Urey's work on the "cosmic epic," Immanuel Velikovsky's book, *Worlds in Collision* (Macmillan, 1950), drew even more direct connections between cosmogony and human history—dramatically linking mythology and catastrophic planetary encounters—and, much to the chagrin of scientists like Urey, proved more popular than their comparatively conservative speculations about the history of the solar system. See Gordin, *The Pseudoscience Wars*.

34. *New Scientist*, "Urey: He Is Tracing the Genesis of the World," 26.

35. Harold C. Urey, "Cooper Union for the Advancement of Science and Art" (typescript, November 2, 1959), p. 7, box 21, folder 19, HCU.

36. Harold C. Urey, "Religion Faces the Atomic Age" (typescript, February 3, 1958), p. 3, box 141, folder 15, HCU. Urey's "prophet" likely was modeled on the diocese of Christ Church, Oxford. Urey had regular and pleasant exchanges with the Reverend Professor John Lowe, dean of Christ Church, and his canons.

37. Harold C. Urey to Winfield E. Little Jr., December 1, 1962, box 54, folder 9, HCU.

38. Urey, "Autobiography, 1970," 40, HCU.

39. Harold C. Urey, "Evolution vs. Miraculous Creation," n.d., p. 13, box 144, folder 25, HCU.

40. Brian, *The Voice of Genius*, 261.

41. Urey, "Autobiography, 1970," 39, HCU.

42. Harold C. Urey to Harold E. Fey, April 8, 1955, box 35, folder 26, HCU.

43. Urey, "Evolution vs. Miraculous Creation," 13, HCU.

44. Urey, "Autobiography, 1970," 40, HCU.

45. Urey to Fey, April 8, 1955.

46. Urey, "Autobiography, 1970," 40, HCU.

47. Urey to Fey, April 8, 1955.

48. Harold C. Urey, "The Intellectual Revolution [Revision]" (typescript, 1956), p. 13, box 141, folder 12, HCU.

49. Anshen is a neglected twentieth-century intellectual. Short biographical treatments can be found in Wyckoff, "Anshen, Ruth Nanda (b. 1900)"; Giusto, "Ruth Nanda Anshen"; and Teich, "Editing Einstein." The description of her as an intellectual instigator comes from Wyckoff's encyclopedia entry.

50. Teich, "Editing Einstein," 24.

51. Giusto, "Ruth Nanda Anshen," 10.

52. Wyckoff, "Anshen, Ruth Nanda (b. 1900)," 54.

53. Giusto, "Ruth Nanda Anshen," 11.

54. Anshen recounted this discussion with Einstein in Giusto, 11.

55. Teich, "Editing Einstein," 24.

56. Anshen quoted in Teich, 24.

57. Harold C. Urey to Ruth Nanda Anshen, May 21, 1940, box 5, RNA.

58. Harold C. Urey to Ruth Nanda Anshen, January 6, 1941, box 5, RNA.

59. Urey to Anshen, January 6, 1941. This speech is quoted above in chapter 6, as it was also published under the title "Accomplishments and Future of Chemical Physics."

60. Teich, "Editing Einstein," 110.

61. Ruth Nanda Anshen to Harold C. Urey, March 5, 1954, box 5, RNA.

62. Harold C. Urey to Ruth Nanda Anshen, March 16, 1954, box 5, RNA.

63. Ruth Nanda Anshen to Harold C. Urey, April 4, 1954, box 5, RNA.

64. Ruth Nanda Anshen to Harold C. Urey, August 3, 1954, box 5, RNA.

65. Anshen to Urey, August 3, 1954.

66. Ruth Nanda Anshen to Harold C. Urey, March 7, 1955, box 5, RNA.

67. Harold C. Urey, "[Response to Sputnik]," n.d., box 2, folder 1, HCU.

68. Urey, "[Response to Sputnik]," HCU.

69. Urey, "[Response to Sputnik]," HCU.

70. Urey, "Affording the Space Program," 24.

71. R. C. Hall, *Lunar Impact*, 11.

72. Newell, *Beyond the Atmosphere*, 237.

73. Newell, 237.

74. Jastrow, *God and the Astronomers*.

75. Jastrow, *Red Giants and White Dwarfs*, 4.

76. Jastrow, 4.

77. Jastrow, "Exploring the Moon," 46.

78. Jastrow, *Red Giants and White Dwarfs*, 4.

79. Jastrow, "Exploring the Moon," 46. According to a note in Urey's papers,

the two men met on December 4, 1958, and spent "a considerable time in discussion . . . that afternoon and the following day." "[Jastrow Meeting Note]," December 4, 1958, box 47, folder 15, HCU.

80. Urey, *The Planets*, 25.

81. Urey, 97–98.

82. Urey, "Primary and Secondary Objects."

83. Jastrow, "Exploring the Moon," 46.

84. Jastrow, 47.

85. Jastrow, 47.

86. Memo quoted in Jastrow, 47.

87. Jastrow, 48.

88. R. C. Hall, *Lunar Impact*, 5.

89. R. C. Hall, 6. ARPA would ultimately become DARPA.

90. Newell, "Harold Urey and the Moon," 1.

91. Jastrow, "Exploring the Moon," 49.

92. Newell, "Harold Urey and the Moon," 1–2.

93. Harold C. Urey to Yale University Press, September 16, 1959, box 105, folder 16, HCU.

94. Harold C. Urey to Robert Jastrow, June 10, 1959, box 47, folder 15, HCU.

95. Harold C. Urey to Robert Jastrow, December 12, 1962, box 47, folder 15, HCU.

96. Harold C. Urey to Al R. Hibbs, July 3, 1959, box 43, folder 3, HCU.

97. Urey to Hibbs, July 3, 1959.

98. Harold C. Urey to Al R. Hibbs, August 10, 1959, box 43, folder 3, HCU.

99. Harold C. Urey to Al R. Hibbs, November 16, 1959, box 43, folder 3, HCU.

100. Memo quoted in Jastrow, "Exploring the Moon," 47.

101. National Aeronautics and Space Administration (NASA), "Press Conference on Lunar Science" (Washington, DC, December 1, 1959), HCU NASA.

102. NASA, "Press Conference on Lunar Science," HCU NASA.

103. This fits with McCurdy's description of efforts by space exploration's early promoters to tap into nostalgia for the bygone days of polar expeditions. See McCurdy, *Space and the American Imagination*, 27–32.

104. Harold C. Urey to Albert Whitford, February 1, 1960, box 51, folder 7, HCU; Harold C. Urey to A. B. Meinel, February 1, 1960, box 51, folder 7, HCU.

105. Whitaker, *The University of Arizona's Lunar and Planetary Laboratory*, 16.

106. Whitaker, 18.

107. Harold C. Urey to Edwin F. Carpenter, January 19, 1960, box 51, folder 7, HCU.

108. Whitaker, *The University of Arizona's Lunar and Planetary Laboratory*, 18.

109. Whitaker, 25.

110. Harold C. Urey to Jan H. Oort, November 9, 1962, box 73, folder 7, HCU.

111. Urey to Oort, November 9, 1962.

112. Urey to Oort, November 9, 1962.

113. Harold C. Urey to Yale University Press, February 3, 1960, box 105, folder 16, HCU.

114. Urey to Yale University Press, February 3, 1960.

115. Harold C. Urey to Yale University Press, January 26, 1961, box 105, folder 16, HCU.

116. Urey to Jastrow, December 12, 1962.

117. Harold C. Urey to Robert Jastrow, June 28, 1963, box 47, folder 15, HCU.

118. Urey, interview by Groueff.

119. Webb quoted in DeVorkin, "Scientists and the Space Sciences," 97.

120. DeVorkin, 102.

121. Urey, "The Contending Moons," 37. Urey repeated this observation so often that the tendency "of almost any theorist to explain a new observation in a way that supports his particular theory" became known among planetary scientists as "Urey's Law": French, *The Moon Book*, 60.

122. Wilhelms, *To a Rocky Moon*, x.

123. A more detailed account of the success of the planetary geologists can be found in Shindell, "Domesticating the Planets."

124. The Lunar Orbiter photos were actually developed and scanned onboard the spacecraft, then returned via video signal to Earth, where the images were regenerated and transferred to 35 mm film. As one might imagine, this technique did produce a number of artifacts in the returned images.

125. See Mutch, *Geology of the Moon*. Mutch's text was written after spending a sabbatical at the USGS Astrogeology Branch, and was completed with the assistance of the USGS astrogeologists.

126. In Mutch's version of the history of geology, the discovery of plate tectonics was not a revolution, but part of this same steady advance of knowledge.

127. Harold C. Urey to Al R. Hibbs, September 26, 1961, box 43, folder 3, HCU.

128. Harold C. Urey to George E. Mueller, October 7, 1969, HCU NASA.

129. Urey, "Autobiography, 1970," 24, HCU.

130. Urey, 24.

131. Urey, 24.

132. Urey, 23.

133. Harold C. Urey, tape 8, interview by Ian I. Mitroff, n.d., p. 6, box 5, folder 20, IIM.

134. Harold C. Urey to Jay Holmes, January 10, 1964, HCU NASA.

135. Director, Apollo Lunar Exploration, to associate administrator for Manned Space Flight, "Comments on Dr. Urey's Letter of Oct 7, 1969," October 31, 1969, HCU NASA.

136. For more on the success of the astrogeologists within NASA corridors, see Beattie, *Taking Science to the Moon*.

137. Born, "Blessings and Evils of Space Travel," 14.

138. Born, 14.

139. Urey, "Affording the Space Program," 25.

140. Urey's letter to O'Keefe is quoted in Brush, "Harold Urey and the Origin of the Moon," 453.

141. Urey, "Statement before Committee of the House of Representatives," 11–12, HCU NASA.

142. *Denver Post*, "Urey Cautions: Moon May Be Disappointing" (news clipping), February 10, 1965, HCU NASA.

143. Urey, "Statement before Committee of the House of Representatives," 12, HCU NASA.

144. Harold C. Urey, handwritten inscription for President John F. Kennedy, February 3, 1961, box 49, folder 8, HCU.

145. Harold C. Urey to Jacqueline Kennedy, December 4, 1963, box 49, folder 8, HCU.

146. Harold C. Urey to Linus Pauling, April 12, 1968, box 419.1, folder 1, LP.

147. Harold C. Urey to Ruth Nanda Anshen, January 5, 1967, box 105, folder 3, HCU.

148. Victor Cohn, "Lunar Study Vindicates Urey," *Washington Post*, August 29, 1969, HCU NASA.

149. Newell, "Harold Urey and the Moon," 5.

150. Harold C. Urey to Douglas Allan, September 21, 1976, box 2, folder 33, HCU.

151. The emergence of the post-Apollo consensus on the Moon's origins is described in Brush, "A History of Modern Selenogony"; see also Brush, *Fruitful Encounters*.

EPILOGUE

1. Harold Lasswell, "Notes on a Visit to Christian Abrahamsen's Studio," box 2, folder 6, HCU; Harold C. Urey to Christian Abrahamsen, May 3, 1979, box 2, folder 6, HCU.

2. Harold C. Urey to Raymond T. Birge, July 13, 1977, box 14, folder 1, HCU.

3. Stevens, "Harold Urey."

4. Frieda Daum Urey to Henry Thode, January 9, 1981, box 90, folder 14, HCU.

5. Harold C. Urey to Raymond T. Birge, March 12, 1979, box 14, folder 1, HCU.

6. Harold C. Urey to Raymond T. Birge, March 10, 1978, box 14, folder 1, HCU.

7. Harold C. Urey, "War and Peace," June 4, 1970, p. 14, box 144, folder 12, HCU.

8. These descriptions of Urey's gravesite are from an obituary titled "Nobel

Winner Buried in Humble Cemetery," reproduced in Reinoehl and Phillips, *Ancestors and Descendants of Solomon and Martha Reinoehl*, 59.

9. Joravsky, *The Lysenko Affair*; Medvedev, *The Rise and Fall of T. D. Lysenko*.

10. Thomas Jukes to Donald DeVincenzi, February 15, 1989, box 160, folder 6, SM.

11. Stanley L. Miller to Thomas Jukes, December 14, 1988, box 160, folder 6, SM.

12. Clayton, "Isotopes."

Archives

ADA: [Alpha Delta Alpha] Delta Omicron Chapter Records. Kappa Sigma Fraternity. Archives and Special Collections, Maureen and Mike Mansfield Library, University of Montana.

CSPR: Conference on Science, Philosophy, and Religion, 1939–1977, record group 5. Special Collections Reading Room, Jewish Theological Seminary Library, Jewish Theological Seminary, New York.

ECAS: Emergency Committee of Atomic Scientists Records, 1946–1952. Special Collections Research Center, University of Chicago Library.

GNL: Gilbert N. Lewis Papers. Bancroft Library and Archives, University of California, Berkeley.

HCU: Harold Clayton Urey Papers, 1929–1981, MSS 44. Mandeville Special Collections Library, Geisel Library, University of California, San Diego, La Jolla.

HCU INS: Harold C. Urey Papers. Institute for Nuclear Studies. Special Collections Research Center, University of Chicago Library.

HCU NASA: Harold C. Urey Files. Archives, History Division, National Aeronautics and Space Administration, Washington, DC.

IIM: Ian I. Mitroff interviews with lunar scientists, 1969–1972. Niels Bohr Library and Archives, American Institute of Physics, College Park, MD.

JTSA: General Files, 1902–1972, record group 1. Special Collections Reading Room, Jewish Theological Seminary of America, New York.

LP: Ava Helen and Linus Pauling Papers, 1901–1994. Special Collections Library, Oregon State University, Corvallis.

MCU: Martha Cullen (Urey) papers. Urey family, private collection.

MJE: Morton J. Elrod Papers, 1885–1959, MSS 486. Archives and Special Collections, Maureen and Mike Mansfield Library, University of Montana, Missoula.

RMH: [Robert Maynard] Hutchins Administration Records. University of Chicago, Office of the President. Special Collections Research Center, University of Chicago Library.

RNA: Ruth Nanda Anshen Papers, 1938–1986, MS 35. Rare Book and Manuscript Library, Columbia University, New York.

RPI: Institute Archives and Special Collections. Rensselaer Libraries, Rensselaer Polytechnic Institute, Troy, NY.

RTB: Raymond T. Birge Papers. Bancroft Library and Archives, University of California, Berkeley.

SE: Samuel Epstein Papers, 10159-MS. Caltech Archives, California Institute of Technology, Pasadena.

SKA: Samuel King Allison Papers, 1920–1965. Special Collections Research Center, University of Chicago Library.

SM: Stanley Miller Papers, 1952–2010, MSS 642. Mandeville Special Collections Library, Geisel Library, University of California, San Diego, La Jolla.

WHS: Walkerton Historical Society Archives, Walkerton, IN.

Oral History Interviews

Mildred Cohn. Interview by Leon Gortler, University of Pennsylvania, December 15, 1987, and January 6, 1988. Chemical Heritage Foundation, Oral History Transcript no. 0080.

General Leslie R. Groves. Interview by Stephane Groueff, January 5–8, 1965. Atomic Heritage Foundation, Voices of the Manhattan Project.

General Kenneth D. Nichols. Interview by Stephane Groueff, January 4–5, 1965. Atomic Heritage Foundation, Voices of the Manhattan Project.

Sir Hugh Taylor. Interview by Stephane Groueff, November 4, 1964. Atomic Heritage Foundation, Voices of the Manhattan Project.

Harold C. Urey. Interview by Harriet Zuckerman, August 26, 1963. "Reminiscences of Harold Clayton Urey: Oral History, 1964," Columbia Center for Oral History, Columbia University.

Harold C. Urey. Interview by John L. Heilbron, March 24, 1964. Niels Bohr Library and Archives, American Institute of Physics, College Park, MD.

Harold C. Urey. Interview by Stephane Groueff, March 3, 1965. Atomic Heritage Foundation, Voices of the Manhattan Project.

John C. Warner. Interview by John A. Heitmann, Gibsonia, Pennsylvania, February 8, 1984. Chemical Heritage Foundation, Oral History Transcript no. 0044.

Bibliography

Aaserud, Finn. *Redirecting Science: Niels Bohr, Philanthropy, and the Rise of Nuclear Physics*. New York: Cambridge University Press, 2003.

Acheson, Dean, and David Lilienthal. *A Report on the International Control of Atomic Energy*. Department of State, Publication 2498. Washington, DC: US Government Printing Office, 1946.

Allison, Samuel K. "Institute for Nuclear Studies, the University of Chicago." *Scientific Monthly* 65, no. 6 (December 1947): 482–88.

Arnold, James R., Jacob Bigeleisen, and Clyde A. Hutchison Jr. "Harold Clayton Urey: April 29, 1893–January 5, 1981." *Biographical Memoirs of the National Academies of Science* 68 (1995): 363–412.

Aronova, Elena. "The Congress for Cultural Freedom, Minerva, and the Quest for Instituting 'Science Studies' in the Age of Cold War." *Minerva* 50 (2012): 307–37.

———. "Studies of Science before 'Science Studies': Cold War and the Politics of Science in the U.S., U.K., and U.S.S.R., 1950s–1970s." PhD diss., University of California, 2012.

Aronova, Elena, and Simone Turchetti, eds. *Science Studies during the Cold War and Beyond: Paradigms Defected*. New York: Palgrave Macmillan, 2016.

Aston, F W. "Bakerian Lecture. A New Mass-Spectrograph and the Whole Number Rule." *Proceedings of the Royal Society of London. Series A, Containing Papers of a Mathematical and Physical Character* 115, no. 772 (1927): 487.

Baldwin, Ralph B. *The Face of the Moon*. Chicago: University of Chicago Press, 1949.

Banner, Lois W. "Biography as History." *American Historical Review* 114, no. 3 (June 2009): 579–86.

Baratta, Joseph Preston. *The Politics of World Federation: From World Federalism to Global Governance*. Greenwood Publishing Group, 2004.

Baxter, Gregory Paul, and Richard Henry Jesse Jr. "A Revision of the Atomic Weight of Chromium: The Analysis of Silver Dichromate." In *Researches upon the Atomic Weights of Cadmium, Manganese, Bromine, Lead, Arsenic, Iodine, Silver, Chromium, and Phosphorus*, 153–64. Washington, DC: Carnegie Institution of Washington, 1910.

Beattie, Donald A. *Taking Science to the Moon: Lunar Experiments and the Apollo Program*. Baltimore: Johns Hopkins University Press, 2001.

Bernstein, Jeremy. "Profiles (I. I. Rabi—I)." *New Yorker*, October 13, 1975.

Birge, Raymond T. *History of the Physics Department, University of California, Berkeley: Volume I. The First Half Century, 1868–1918*. Berkeley: University of California, 1968.

———. *History of the Physics Department, University of California, Berkeley: Volume II. The Decade 1918–1928*. Berkeley: University of California, 1968.

Black, Edwin. *War against the Weak: Eugenics and America's Campaign to Create a Master Race*. New York: Four Walls Eight Windows, 2003.

Bohr, Niels, and Finn Aaserud. *Popularization and People (1911–1962)*. Amsterdam: Elsevier, 2007.

Born, Max. "Blessings and Evils of Space Travel." *Bulletin of the Atomic Scientists* 22, no. 8 (October 1966): 12–14.

Bowman, Carl F., and Carl Desportes Bowman. *Brethren Society: The Cultural Transformation of a "Peculiar People."* Baltimore: Johns Hopkins University Press, 1995.

Bowman, Rufus David. *The Church of the Brethren and War, 1708–1941*. Elgin, IL: Brethren Publishing House, 1944.

Boyer, Paul. *By the Bomb's Early Light: American Thought and Culture at the Dawn of the Atomic Age*. New York: Pantheon, 1985.

Brian, Denis. *The Voice of Genius: Conversations with Nobel Scientists and Other Luminaries*. Cambridge, MA: Perseus, 2001.

Brickwedde, Ferdinand G. "Atomic Models as Proposed by Bohr." Master's thesis, Johns Hopkins University, 1924.

———. "Harold Urey and the Discovery of Deuterium." *Physics Today*, September 1982.

Brooke, John Hedley. "Religious Belief and the Content of the Sciences." In *Science in Theistic Contexts: Cognitive Dimensions*, Osiris, vol. 16, edited by Brooke,

Margaret J. Osler, and Jitse M. van der Meer, 3–28. Chicago: University of Chicago Press.

———. *Science and Religion: Some Historical Perspectives*. New York: Cambridge University Press, 1991.

Brooke, John Hedley, Margaret J. Osler, and Jitse M. van der Meer, eds. *Science in Theistic Contexts: Cognitive Dimensions*. Osiris, vol. 16. Chicago: University of Chicago Press, 2001.

Brush, Stephen G. *Fruitful Encounters: The Origin of the Solar System and of the Moon from Chamberlain to Apollo*. A History of Modern Planetary Physics 3. New York: Cambridge University Press, 1996.

———. "Harold Urey and the Origin of the Moon: The Interaction of Science and the Apollo Program." In *Spacelab, Space Platforms, and the Future*, edited by Peter Bainum and Dietrich E. Koelle, 49:437–70. Advances in the Astronautical Sciences. San Diego: American Astronautical Society, 1982.

———. "A History of Modern Selenogony: Theoretical Origins of the Moon, from Capture to Crash, 1955–1984." *Space Science Reviews* 47 (1988): 211–73.

———. "Theories of the Origin of the Solar System 1956–1985." *Reviews of Modern Physics* 62, no. 1 (January 1, 1990): 43.

Bryson, Lyman, and Louis Finkelstein, eds. *Science, Philosophy, and Religion. Second Symposium*. New York: Conference on Science, Philosophy, and Religion in Their Relation to the Democratic Way of Life, 1942.

Buettler, Fred W. "Organizing an American Conscience: The Conference on Science, Philosophy and Religion, 1940–1968." PhD diss., University of Chicago, 1995.

Calvin, Melvin. *Gilbert Newton Lewis: His Influence on Physical-Organic Chemists at Berkeley*. Berkeley: Lawrence Berkeley National Laboratory, 1982.

———. "Gilbert Newton Lewis: His Influence on Physical-Organic Chemists at Berkeley." *Journal of Chemical Education* 61, no. 1 (January 1984): 14–18.

Carnegie Institution of Washington. *Year Book No. 7, 1908*. Washington, DC: Carnegie Institution of Washington, 1909.

Chemical Engineer. "The American Chemist Must Enlist." Vol. 25, no. 2 (April 1917): 41.

Clayton, Robert N. "Isotopes: From Earth to the Solar System." *Annual Review of Earth and Planetary Sciences* 35, no. 1 (2007): 1–19.

Coben, Stanley. "The Scientific Establishment and the Transmission of Quantum Mechanics to the United States, 1919–32." *American Historical Review* 76, no. 2 (1971): 442–66.

Coffey, Patrick. *Cathedrals of Science: The Personalities and Rivalries That Made Modern Chemistry*. New York: Oxford University Press, 2008.

———. "Chemical Free Energies and the Third Law of Thermodynamics." *Historical Studies in the Physical and Biological Sciences* 36, no. 2 (March 1, 2006): 365–96.

Cohen, Karl. "Harold C. Urey 1893–1981." *Bulletin of the Atomic Scientists* 37, no. 5 (May 1981): 8–56.

Cohen, Karl P., S. Keith Runcorn, Hans E. Suess, and Harry G. Thode. "Harold Clayton Urey. 29 April 1893–5 January 1981." *Biographical Memoirs of Fellows of the Royal Society* 29 (November 1983): 623–59.

Cohen-Cole, Jamie. *The Open Mind: Cold War Politics and the Sciences of Human Nature*. Chicago: University of Chicago Press, 2014.

Cohn, Mildred. "Harold Urey: A Personal Remembrance, Part II." *Chemical Heritage Magazine*, Spring 2006.

Committee to Frame a World Constitution. *Preliminary Draft of a World Constitution*. Chicago: University of Chicago Press, 1948.

Compton, Karl T. "Science Still Holds Great Promise." *New York Times*, December 16, 1934.

Craig, Harmon, Stanley L. Miller, and Gerald J. Wasserburg, eds. *Isotopic and Cosmic Chemistry*. Amsterdam: North-Holland Publishing, 1964.

Creager, Angela N. H. "Atomic Tracings: Radioisotopes in Biology and Medicine." In *Science and Technology in the Global Cold War*, edited by Naomi Oreskes and John Krige, 31–73. Cambridge, MA: MIT Press, 2014.

———. *Life Atomic: A History of Radioisotopes in Science and Medicine*. Chicago: University of Chicago Press, 2013.

Cruikshank, Dale P. "Gerard Peter Kuiper, December 7, 1905–December 24, 1973." In *Biographical Memoirs*, 62:259–95. Washington, DC: National Academies Press, 1993.

Cruikshank, Dale P., George Coyne, and Alan Binder. "Gerard P. Kuiper." Accessed September 15, 2016. https://www.lpl.arizona.edu/history/points-of-light/founding/kuiper.

Davenport, Charles Benedict. *Heredity in Relation to Eugenics*. New York: H. Holt, 1911.

Davies, Shannon M. "American Physicists Abroad: Copenhagen, 1920–1940." PhD diss., University of Texas at Austin, 1985.

Davis, Edward B. "Prophet of Science—Part One: Arthur Holly Compton on Science, Freedom, Religion, and Morality." *Perspectives on Science and Christian Faith* 61, no. 2 (June 2009): 73–83.

———. "Prophet of Science—Part Two." *Perspectives on Science and Christian Faith* 61, no. 3 (September 2009): 175–90.

———. "Prophet of Science—Part Three." *Perspectives on Science and Christian Faith* 61, no. 4 (December 2009): 240–53.

———. "Robert Andrews Millikan: Religion, Science, and Modernity." In *Eminent Lives in Twentieth-Century Science & Religion*, edited by Nicolaas A. Rupke, 253–74. Frankfurt: Peter Lang, 2009.

Dawsey, Lynn H. "The Absorption Spectrum and Photochemical Dissociation of Hydrogen Peroxyde." PhD diss., Johns Hopkins University, 1929.

DeMaggio, Susan. "Chronicling Qualcomm: Irwin Jacobs Lectures at the Riford Center." *La Jolla Light*. July 27, 2011.

Dennis, Michael Aaron. "Historiography of Science: An American Approach." In *Science in the Twentieth Century*, edited by John Krige and Dominique Pestre, 1–26. Amsterdam: Harwood Academic Publishers, 1997.

DeVorkin, David H. "Scientists and the Space Sciences." In *Space: Discovery and Exploration*, edited by Martin Collins and Sylvia Kraemer, 69–111. Washington, DC: Smithsonian National Air and Space Museum, 1993.

Dick, Steven J., and James E. Strick. *The Living Universe: NASA and the Development of Astrobiology*. New Brunswick: Rutgers University Press, 2004.

Doel, Ronald E. "The Earth Sciences and Geophysics." In *Companion to Science in the Twentieth Century*, edited by John Krige and Dominique Pestre, 391–416. New York: Routledge, 2003.

———. *Solar System Astronomy in America: Communities, Patronage, and Interdisciplinary Science, 1920–1960*. New York: Cambridge University Press, 1996.

Doel, Ronald E., Tanya J. Levin, and Mason K. Marker. "Extending Modern Cartography to the Ocean Depths: Military Patronage, Cold War Priorities, and the Heezen–Tharp Mapping Project, 1952–1959." *Journal of Historical Geography* 32, no. 3 (July 2006): 605–26.

Dove, Frederick Denton. "Cultural Changes in the Church of the Brethren: A Study in Cultural Sociology." PhD diss., University of Pennsylvania, 1932.

Eckel, George. "Red Inquiry Hears Attack on Dr. Urey." *New York Times*. April 24, 1949.

Edwards, Morgan. *Materials towards a History of the American Baptists*. Philadelphia: Joseph Crukshank and Isaac Collins, 1770.

E. F. A. Review of *Inorganic Chemistry*, by Fritz Ephraim and P. C. L. Thorne. *Journal of the Royal Society of Arts* 82, no. 4240 (1934): 430–32.

Epstein, Samuel, Ralph Buchsbaum, Heinz A. Lowenstam, and Harold C. Urey.

"Carbonate-Water Isotopic Temperature Scale." *Bulletin of the Geological Society of America* 62 (April 1951): 417–26.

Erdmann, Karl Dietrich. *Toward a Global Community of Historians: The International Historical Congresses and the International Committee of Historical Sciences, 1898–2000*. New York: Berghahn Books, 2005.

Eshelman, M. M., W. E. Trostle, W. H. Keim, D. A. Norcross, and B. F. Masterson. *A History of the Church of the Brethren of Southern California and Arizona*. Los Angeles: District Meeting of Southern California and Arizona, 1917.

Eyring, Henry. "The New Point of View in Chemistry." *Scientific Monthly* 39, no. 5 (1934): 415–19.

Fermi, Laura. *Atoms in the Family: My Life with Enrico Fermi*. Chicago: University of Chicago Press, 1995.

Fisher, Daniel. *The Seventy Years of the Department of Geology, University of Chicago, 1892–1961*. Chicago: University of Chicago, 1963.

Flynn, George Q. "The Draft and College Deferments during the Korean War." *Historian* 50, no. 3 (May 1988): 369–85.

Forman, Paul. "Weimar Culture, Causality, and Quantum Theory, 1918–1927: Adaptation by German Physicists and Mathematicians to a Hostile Intellectual Environment." *Historical Studies in the Physical Sciences* 3 (January 1971): 1–115.

French, Bevan M. *The Moon Book*. New York: Penguin Books, 1977.

Garner, Steve. *Whiteness*. New York: Routledge, 2007.

Garrison, W. M., D. C. Morrison, J. G. Hamilton, A. A. Benson, and M. Calvin. "Reduction of Carbon Dioxide in Aqueous Solutions by Ionizing Radiation." *Science*, n.s., 114, no. 2964 (October 19, 1951): 416–18.

Gellhorn, Walter. *Security, Loyalty, and Science*. Ithaca: Cornell University Press, 1950.

Giauque, William F., and Helen L. Johnston. "The Heat Capacity of Oxygen from 12°K to Its Boiling Point and Its Heat of Vaporization." *Journal of the American Chemical Society* 51, no. 8 (August 1929): 2300–2321.

———. "An Isotope of Oxygen, Mass 17, in the Earth's Atmosphere." *Journal of the American Chemical Society* 51, no. 12 (December 1929): 3528–34.

———. "An Isotope of Oxygen, Mass 18. Interpretation of the Atmospheric Absorption Bands." *Journal of the American Chemical Society* 51, no. 5 (May 1929): 1436–41.

Giusto, Joann. "Ruth Nanda Anshen." *Publishers Weekly*, January 9, 1978.

Goldsmith, Julian R. "Some Chicago Georecollections." *Annual Review of Earth and Planetary Sciences* 19 (1991): 1–16.

Gordin, Michael D. *The Pseudoscience Wars: Immanuel Velikovsky and the Birth of the Modern Fringe*. Chicago: University of Chicago Press, 2012.

———. *A Well-Ordered Thing: Dmitrii Mendeleev and the Shadow of the Periodic Table*. New York: Basic Books, 2004.

Grayson, Michael A. *Measuring Mass: From Positive Rays to Proteins*. Philadelphia: Chemical Heritage Press, 2002.

Greenberg, Daniel S. *The Politics of Pure Science*. New York: New American Library, 1971.

———. *Science, Money, and Politics: Political Triumph and Ethical Erosion*. Chicago: University of Chicago Press, 2001.

Groueff, Stephane. *Manhattan Project: The Untold Story of the Making of the Atomic Bomb*. Boston: Little, Brown, 1967.

Hall, Mary Harrington. "All Sides of Harold Urey." *San Diego*, November 1960.

Hall, Mary Harrington, and Harold C. Urey. "As I See It." *Forbes*, July 15, 1969.

Hall, R. Cargill. *Lunar Impact: A History of Project Ranger*. NASA History Series 4210. Washington, DC: National Aeronautics and Space Administration, 1977.

Harrington, John Walker. "American Progress in Chemical Arts." *New York Times*, September 29, 1918.

Hawkins, Helen S., G. Allen Greb, and Gertrud Weiss Szilard, eds. *Toward a Livable World: Leo Szilard and the Crusade for Nuclear Arms Control*. The Collected Works of Leo Szilard, vol. 3. Cambridge, MA: MIT Press, 1987.

Heilbron, John L., and Robert W. Seidel. *Lawrence and His Laboratory: A History of the Lawrence Berkeley Laboratory*. California Studies in the History of Science. Berkeley: University of California Press, 1989.

Hershberg, James G. *James B. Conant: Harvard to Hiroshima and the Making of the Nuclear Age*. Stanford: Stanford University Press, 1995.

Hewlett, Richard G., and Oscar E. Anderson, Jr. *The New World, 1939–1946*. A History of the United States Atomic Energy Commission, vol. 1. University Park: Pennsylvania State University Press, 1962.

Hildebrand, Joel H. "The Air That Harold C. Urey Breathed in Berkeley." In *Isotopic and Cosmic Chemistry*, edited by Harmon Craig, Stanley L. Miller, and Gerald J. Wasserburg, viii–ix. Amsterdam: North-Holland Publishing, 1963.

———. "Gilbert Newton Lewis, 1875–1946." *Obituary Notices of Fellows of the Royal Society* 5, no. 15 (February 1947): 491–506.

Hogan, Herbert W. "The Intellectual Impact of the Twentieth Century on the Church of the Brethren." PhD diss., Claremont Graduate School, 1958.

Hollinger, David A. "The Defense of Democracy and Robert K. Merton's Formulation of the Scientific Ethos." *Knowledge and Society* 4 (1983): 1–15.

———. "Science as a Weapon in Kulturkampfe in the United States during and after World War II." *Isis* 86, no. 3 (1995): 440–54.

———. *Science, Jews, and Secular Culture: Studies in Mid-Twentieth-Century American Intellectual History*. Princeton, NJ: Princeton University Press, 1998.

Housholder, Terry. "Kendallville Graduate Worked on Manhattan Project in World War II." *Kendallville News-Sun*, November 4, 1999. Reflections of the Century, 1900–1999, special edition, part 3, sec. C and D. Last accessed August 29, 2011. http://www.kpcnews.net/special-sections/reflections3/index.html.

Hutchins, Robert M. "The Man We Love." In *Isotopic and Cosmic Chemistry*, edited by Harmon Craig, Stanley L. Miller, and Gerald J. Wasserburg, v–vii. Amsterdam: North-Holland Publishing, 1964.

Hutchison, Clyde A., Jr. "Harold Clayton Urey, 1893–1981." In *Remembering the University of Chicago: Teachers, Scientists, and Scholars*, edited by Edward Shils, 515–32. Chicago: University of Chicago Press, 1991.

Ingersoll, Robert G. *About the Holy Bible: A Lecture*. New York: C. P. Farrell, 1894.

Jacobson, Matthew Frye. *Whiteness of a Different Color: European Immigrants and the Alchemy of Race*. Cambridge, MA: Harvard University Press, 1998.

Jacoby, Susan. *Freethinkers: A History of American Secularism*. New York: Metropolitan Books, 2004.

Jastrow, Robert. "Exploring the Moon." In *Space Science Comes of Age*, edited by Paul A. Hanle and Von Del Chamberlain, 45–50. Washington, DC: National Air and Space Museum, 1981.

———. *God and the Astronomers*. New York: W. W. Norton, 1978.

———. *Red Giants and White Dwarfs*. New York: W. W. Norton, 1967.

Jolly, William L. *From Retorts to Lasers: The Story of Chemistry at Berkeley*. Berkeley, CA: College of Chemistry, 1987.

Joravsky, David. *The Lysenko Affair*. Cambridge, MA: Harvard University Press, 1970.

Journal of Industrial and Engineering Chemistry. "Chemical Warfare Service." Vol. 10, no. 9 (September 1918): 675–84.

———. "The Parting of the Ways." Vol. 10, no. 4 (April 1, 1918): 254–55.

———. "Personal Notes." Vol. 12, no. 1 (1920): 93–95.

Kevles, Daniel. *The Physicists: The History of a Scientific Community in Modern America*. Cambridge, MA: Harvard University Press, 2001.

———. "The Study of Physics in America, 1865–1916." PhD diss., Princeton University, 1964.

Kohler, Robert E., Jr. "Rudolf Schoenheimer, Isotopic Tracers, and Biochemistry in the 1930's." *Historical Studies in the Physical Sciences* 8 (1977): 257–98.

Krige, John, and Dominique Pestre, eds. *Science in the Twentieth Century*. Amsterdam: Harwood Academic Publishers, 1997.

Kuznick, Peter J. *Beyond the Laboratory: Scientists as Political Activists in 1930s America*. Chicago: University of Chicago Press, 1987.

Lang, Daniel. "That's Four Times 10^{-4} Ergs, Old Man." *New Yorker*, November 16, 1946.

Lavin, George Israel. "Reactions of Atomic Hydrogen and of Dissociated Water Vapor." PhD diss., Johns Hopkins University, 1928.

Lehman, James O., and Steven M. Nolt. *Mennonites, Amish, and the American Civil War*. Baltimore: Johns Hopkins University Press, 2007.

Leslie, Stuart W. *The Cold War and American Science: The Military-Industrial-Academic Complex at MIT and Stanford*. New York: Columbia University Press, 1993.

Lewis, Gilbert N., and Ronald T. MacDonald. "Concentration of H2 Isotope." *Journal of Chemical Physics* 1 (June 1933): 341–44.

Lindberg, David C., and Ronald L. Numbers, eds. *God and Nature: Historical Essays on the Encounter between Christianity and Science*. Berkeley: University of California Press, 1986.

———, eds. *When Science & Christianity Meet*. Chicago: University of Chicago Press, 2003.

Los Angeles Herald (Los Angeles, California), "A Bad Case, A Minister of the Brethren Church Is Sent to Highland," July 10, 1897.

Lowell, A. Lawrence. *Reports of the President and the Treasurer of Harvard College, 1917–1918*. Cambridge, MA: Harvard University, 1919.

Lowen, Rebecca S. *Creating the Cold War University: The Transformation of Stanford*. Berkeley: University of California Press, 1997.

Luebke, Frederick C. *Bonds of Loyalty: German-Americans and World War I*. DeKalb: Northern Illinois University Press, 1974.

Mack, Pamela E. "Strategies and Compromises: Women in Astronomy at Harvard College Observatory, 1870–1920." *Journal for the History of Astronomy* 21, no. 1 (1990): 65–76.

Malone, Michael P., Richard B. Roeder, and William L. Lang. *Montana: A History of Two Centuries*. Seattle: University of Washington Press, 1976.

Marble, John P. "In Memoriam, Gregory Paul Baxter, 1876–1953." In *Report of the Committee on the Measurement of Geologic Time*, i. Washington: National Academy of Sciences, 1953.

McCaughey, Robert. *Stand, Columbia: A History of Columbia University*. Columbia University Press, 2012.

McCurdy, Howard E. *Space and the American Imagination*. Baltimore: Johns Hopkins University Press, 2011.

McGrayne, Sharon Bertsch. *Nobel Prize Women in Science: Their Lives, Struggles, and Momentous Discoveries*. Washington, DC: Joseph Henry Press, 2006.

Medvedev, Zhores A. *The Rise and Fall of T. D. Lysenko*. New York: Columbia University Press, 1969.

Meitner, Lise, and Otto R. Frisch. "Disintegration of Uranium by Neutrons: A New Type of Nuclear Reaction." *Nature* 143, no. 3615 (February 11, 1939): 239–40.

Milam, Erika L., and Robert A. Nye, eds. *Scientific Masculinities*. Osiris, vol. 30. Chicago: University of Chicago Press, 2015.

Miller, Stanley L. "A Production of Amino Acids under Possible Primitive Earth Conditions." *Science* 117, no. 3046 (May 15, 1953): 528–29.

Mitroff, Ian I. "Norms and Counter-norms in a Select Group of the Apollo Moon Scientists: A Case Study of the Ambivalence of Scientists." *American Sociological Review* 39, no. 4 (1974): 579.

———. "On Evaluating the Scientific Contribution of the Apollo Moon Missions via Information Theory: A Study of the Scientist-Scientist Relationship." *Management Science* 20, no. 12 (1974): 1501.

———. *The Subjective Side of Science: A Philosophical Inquiry into the Psychology of the Apollo Moon Scientists*. New York: American Elsevier, 1974.

Mukerji, Chandra. *A Fragile Power: Scientists and the State*. Princeton, NJ: Princeton University Press, 1989.

Mullaney, Thomas E. "Atomic Research Shows Progress." *New York Times*, June 26, 1949.

Murphy, George M., and Harold C. Urey. "On the Relative Abundances of the Nitrogen and Oxygen Isotopes." *Physical Review* 41, no. 2 (July 15, 1932): 141.

Mutch, Thomas A. *Geology of the Moon: A Stratigraphic View*. Princeton, NJ: Princeton University Press, 1970.

Nasaw, David. "Introduction to AHR Roundtable: Historians and Biography." *American Historical Review* 114, no. 3 (June 2009): 573–78.

National Academy of Sciences. "Abstracts of Papers Presented at the 1949 Autumn Meeting, Rochester, New York." *Science* 110, no. 2861 (October 28, 1949): 437–46.

Needell, Allan. *Science, Cold War and the American State: Lloyd V. Berkner and the Balance of Professional Ideals*. Amsterdam: Harwood Academic Publishers, 2000.

Newell, Homer E. *Beyond the Atmosphere: Early Years of Space Science*. Washington, DC: National Aeronautics and Space Administration, 1980.

———. "Harold Urey and the Moon." *Moon* 7, no. 1–2 (1973): 1–5.
New Scientist. "Harold C. Urey: He Is Tracing the Genesis of the World." January 17, 1957.
New Yorker. "The Talk of the Town: Notes and Comment." December 15, 1945.
New York Times. "Berle Sees Nazis Losing Strength." September 9, 1941.
———. "Bid Farewell to Urey." February 2, 1935.
———. "Brooklyn Professor, Called a Red, Bars House Queries on Atom Data." May 25, 1949.
———. "Our Place in Science." April 15, 1934.
———. "Scientific Romancing." October 30, 1949.
———. "Secrecy on Atom Called Excessive." September 25, 1949.
———. "Urey Prefers Shell Work to Atomic Researching." January 25, 1950.
Nier, Alfred O. C. "Some Reminiscences of Mass Spectrometry and the Manhattan Project." *Journal of Chemical Education* 66, no. 5 (May 1989): 385–88.
Nobel Foundation. *Nobel Lectures, Chemistry 1922–1941*. Amsterdam: Elsevier, 1966.
Nye, Mary Jo. *Before Big Science: The Pursuit of Modern Chemistry and Physics, 1800–1940*. Cambridge, MA: Harvard University Press, 1999.
———. *From Chemical Philosophy to Theoretical Chemistry: Dynamics of Matter and Dynamics of Disciplines, 1800–1950*. Berkeley: University of California Press, 1993.
———. "Scientific Biography: History of Science by Another Means?" *Isis* 97, no. 2 (2006): 322–29.
Oreskes, Naomi. *Science on a Mission: American Oceanography in the Cold War and Beyond*. Chicago: University of Chicago Press, forthcoming.
Paul, Diane. *Controlling Human Heredity: 1865 to the Present*. Atlantic Highlands: Humanities Press International, 1995.
Pitzer, Kenneth S., and David A. Shirley. "William Francis Giauque, 1895–1982." *Biographical Memoirs* 70 (1996): 39–57.
Plumb, Robert K. "Earth Heating Up, Urey Theory Holds." *New York Times*, October 27, 1949.
Porter, Theodore M. "Is the Life of the Scientist a Scientific Unit?" *Isis* 97, no. 2 (2006): 314–21.
Rainger, Ronald. "Science at the Crossroads: The Navy, Bikini Atoll, and American Oceanography in the 1940s." *Historical Studies in the Physical and Biological Sciences* 30, no. 2 (January 1, 2000): 349–71.
Randall, Merle. "Gilman Hall: The Research Unit of the Chemistry Group at the University of California." *Journal of Industrial and Engineering Chemistry* 10, no. 8 (August 1918): 634–40.

Reed, Germaine M. *Crusading for Chemistry: The Professional Career of Charles Holmes Herty*. Athens: University of Georgia Press, 1995.

Reinhardt, Carsten. "The Chemistry of an Instrument: Mass Spectrometry and Structural Organic Chemistry." In *From Classical to Modern Chemistry: The Instrumental Revolution*, edited by Peter J. T. Morris, 229–50. Cambridge: Royal Society of Chemistry, 2002.

———. *Shifting and Rearranging: Physical Methods and the Transformation of Modern Chemistry*. Sagamore Beach: Science History Publications, 2006.

Reinoehl, Charles M., and George B. Eckhart. *History of an Eckhar(d)t Family Whose Three Sons (John, Henry, George) Came to America before 1850*. Bryan, TX: Scribe Shop, 1952.

Reinoehl, Charles M., and Jennie King Phillips. *Ancestors and Descendants of Solomon and Martha Reinoehl*. Rev. ed. Self-published, 1995.

Reinoehl, Paul E. *A History of the Fairfield Cemetery: The Most Famous Man Buried Therein, Dr. Harold Clayton Urey, the Fairfield Township Schools, the Hamlet of Fairfield Center*. Ashley, IN: Reinoehl, 1998.

Rhees, David J. "The Chemists' Crusade: The Rise of an Industrial Science in Modern America, 1907–1922." PhD diss., University of Pennsylvania, 1987.

———. "The Chemists' War: The Impact of World War I on the American Chemical Profession." *Bulletin for the History of Chemistry*, no. 13–14 (1992): 40–47.

Rhodes, Richard. *The Making of the Atomic Bomb*. New York: Simon & Schuster, 1986.

Richards, Joan L. "Introduction: Fragmented Lives." *Isis* 97, no. 2 (2006): 302–5.

Richards, Theodore William, and Richard Henry Jesse Jr. "The Heats of Combustion of the Octanes and Xylenes." *Journal of the American Chemical Society* 32, no. 3 (March 1910): 268–98.

Rigden, John S. *Rabi: Scientist and Citizen*. Cambridge, MA: Harvard University Press, 2000.

Robertson, Peter. *The Early Years: The Niels Bohr Institute, 1921–1930*. Copenhagen: Akademisk Forlag, 1979.

Roe, Anne. *The Making of a Scientist*. New York: Dodd Mead, 1953.

Rossiter, Margaret W. *Women Scientists in America: Struggles and Strategies to 1940*. Baltimore: Johns Hopkins University Press, 1992.

Ruark, Arthur E., and Harold C. Urey. *Atoms, Molecules and Quanta*. New York: McGraw-Hill, 1930.

Rubin, Vera. *Bright Galaxies, Dark Matters*. Woodbury: American Institute of Physics, 1997.

Rupke, Nicolaas A., ed. *Eminent Lives in Twentieth-Century Science & Religion*. New York: Peter Lang, 2007.

Rutherford, Lord, N. V. Sidgwick, F. W. Aston, P. Harteck, F. Soddy, M. Polanyi, E. K. Rideal, et al. "Discussion on Heavy Hydrogen: Opening Address." *Proceedings of the Royal Society of London. Series A, Containing Papers of a Mathematical and Physical Character* 144, no. 851 (1934): 1–28.

Sappington, Roger Edwin. *The Brethren in Industrial America: A Source Book on the Development of the Church of the Brethren, 1865–1915*. Elgin: Brethren Press, 1985.

Schoen, Max. "The Basis for Faith in Democracy." In *Science, Philosophy, and Religion. Second Symposium*, edited by Lyman Bryson and Louis Finkelstein, 98–109. New York: Conference on Science, Philosophy, and Religion in Their Relation to the Democratic Way of Life, 1942.

Schweber, Silvan S. "The Young John Clarke Slater and the Development of Quantum Chemistry." *Historical Studies in the Physical and Biological Sciences* 20, no. 2 (January 1990): 339–406.

Science. "Scientists in the News." N.s., 126, no. 3262 (July 5, 1957): 22–23.

Scott, Franklin W., ed. *The Semi-centennial Alumni Record of the University of Illinois*. Urbana-Champaign: University of Illinois, 1918.

Seidel, Robert W. "Physics Research in California: The Rise of a Leading Sector in American Physics." PhD diss., University of California, Berkeley, 1978.

Servos, John W. *Physical Chemistry from Ostwald to Pauling: The Making of a Science in America*. Princeton, NJ: Princeton University Press, 1996.

Shelton, William R. "Harold Urey, Adventurer." In *1965, Science Year: The World Book Science Annual*. Chicago: Field Enterprises Educational Corporation, 1965.

Shenk, Gerald E. *"Work or Fight!": Race, Gender, and the Draft in World War One*. New York: Palgrave Macmillan, 2005.

Shindell, Matthew B. "Domesticating the Planets: Instruments and Practices in the Development of Planetary Geology." *Spontaneous Generations: A Journal for the History and Philosophy of Science* 4, no. 1 (2010): 191–230.

Shortland, Michael, and Richard Yeo, eds. *Telling Lives in Science: Essays on Scientific Biography*. New York: Cambridge University Press, 1996.

Silverstein, Alvin, and Virginia Silverstein. *Harold Urey: The Man Who Explored from Earth to Moon*. Illustrated by Lee J. Ames. New York: J. Day, 1970.

Simpson, Christopher, ed. *Universities and Empire: Money and Politics in the Social Sciences during the Cold War*. New York: New Press, 1998.

Slichter, Louis Byrne. "Cooling of the Earth." *Bulletin of the Geological Society of America* 52, no. 4 (April 1941): 561–600.

Smallwood, Hugh M. Review of *Atoms, Molecules and Quanta*, by Harold C. Urey and Arthur E. Ruark. *Journal of the American Chemical Society* 52, no. 6 (June 1930): 2588–89.

———. "A Study of the Active Modifications of Hydrogen." PhD diss., Johns Hopkins University, 1927.

Smith, Alice Kimball. *A Peril and a Hope: The Scientists' Movement in America, 1945–47*. Chicago: University of Chicago Press, 1965.

Smyth, H. D. "Atomic Energy for Military Purposes." *Reviews of Modern Physics* 17, no. 4 (October 1, 1945): 351–471.

Stanley, Matthew. *Practical Mystic: Religion, Science, and A. S. Eddington*. Chicago: University of Chicago Press, 2007.

State University of Montana (Missoula). *Twenty-First Annual Register: 1916–1917*. Missoula: Bureau of Printing, 1916.

Stevens, Gus. "Harold Urey: A Genius Lives in Failing Body." *Beaver County Times*, May 21, 1979, sec. A.

Streit, Clarence K. *Union Now: A Proposal for a Federal Union of the Democracies of the North Atlantic*. New York: Harper & Bros, 1939.

———. *Union Now with Britain*. New York: Harper & Bros, 1941.

Suess, Hans E., and Harold C. Urey. "Abundances of the Elements." *Reviews of Modern Physics* 28, no. 1 (January 1, 1956): 53.

Talbot, Henry P. "Chemistry at the Front." *Atlantic Monthly*, August 1918.

Teich, Mark. "Editing Einstein." *Omni*, July 1988.

Telegdi, Valentine L. "Enrico Fermi, 1901–1954." In *Remembering the University of Chicago: Teachers, Scientists, and Scholars*, edited by Edward Shils, 110–29. Chicago: University of Chicago Press, 1991.

Terrall, Mary. "Biography as Cultural History of Science." *Isis* 97, no. 2 (2006): 306–13.

Thomas, Shirley. "Harold C. Urey." In *Men of Space: Profiles of Scientists Who Probe for Life in Space*, 6:212–41. Philadelphia: Chilton Books, 1963.

Thorpe, Charles. *Oppenheimer: The Tragic Intellect*. Chicago: University of Chicago Press, 2006.

Tolman, Richard C., and Richard M. Badger. "The Entropy of Diatomic Gases and Rotational Specific Heat." *Journal of the American Chemical Society* 45, no. 10 (October 1, 1923): 2277–85.

Trytten, Merriam Hartwick. *Student Deferment in Selective Service: A Vital Factor in National Security*. Minneapolis: University of Minnesota Press, 1952.

Uhlenbeck, George E., and Samuel A. Goudsmit. "Ersetzung Der Hypothese Vom Unmechanischen Zwang Durch Eine Forderung Bezüglich Des Inneren Verhaltens Jedes Einzelnen Elektrons." *Die Naturwissenschaften* 13, no. 47 (November 1925): 953–54.

University of Montana Junior Class. *The 1915 Sentinel: The Yearbook of the University of Montana*. Vol. 11. Missoula: University of Montana, 1915. https://scholarworks.umt.edu/sentinel/12/.

———. *The 1917 Sentinel: The Yearbook of the University of Montana*. Vol 13. Missoula: University of Montana, 1917. https://scholarworks.umt.edu/sentinel/14/.

Urey, Harold C. "The Abundances of the Elements." *Physical Review* 88, no. 2 (October 15, 1952): 248.

———. "Accomplishments and Future of Chemical Physics." *Review of Scientific Instruments* 8, no. 7 (1937): 223.

———. "Affording the Space Program." *Bulletin of the Atomic Scientists* 23, no. 2 (February 1967): 24–25.

———. "Chemical Fractionation in the Meteorites and the Abundance of the Elements." *Geochimica et Cosmochimica Acta* 2, no. 5–6 (1952): 269–82.

———. "Chemically Pure." *Bulletin of the Atomic Scientists* 11, no. 2 (February 1955): 65, 72.

———. "Chemistry and the Future." *Science* 88, no. 2276 (1938): 133–39.

———. "The Contending Moons." *Astronautics and Aeronautics* 7, no. 1 (January 1969): 37–41.

———. "The Cosmic Abundances of Potassium, Uranium, and Thorium and the Heat Balances of the Earth, the Moon, and Mars." *Proceedings of the National Academy of Sciences of the United States of America* 41, no. 3 (1955): 127.

———. "Diamonds, Meteorites, and the Origin of the Solar System." *Astrophysical Journal* 124 (November 1956): 623.

———. "Discussion." *Journal of Chemical Education* 6, no. 2 (February 1929): 258–59.

———. "The Distribution of Electrons in the Various Orbits of the Hydrogen Atom." *Astrophysical Journal* 59, no. 1 (January 1924): 1–10.

———. "The Heat Capacities and Entropies of Diatomic and Polyatomic Gases." *Journal of the American Chemical Society* 45, no. 6 (June 1923): 1445–55.

———. "On the Early Chemical History of the Earth and the Origin of Life." *Proceedings of the National Academy of Sciences of the United States of America* 38, no. 4 (April 1952): 351–63.

———. "On the Effect of Perturbing Electric Fields on the Zeeman Effect of the

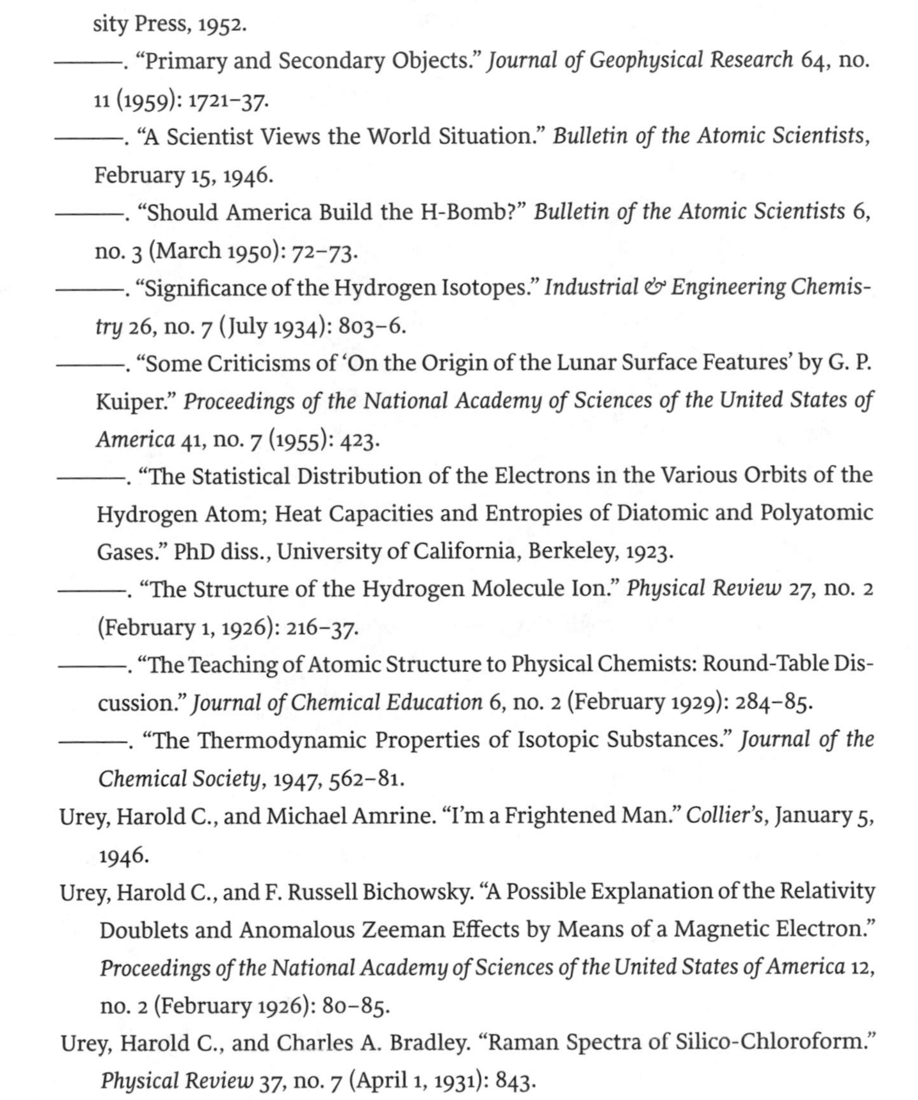

Hydrogen Spectrum." *Kongelige Danske Videnskabernes Selskab, Mathematisk-Fysiske Meddelelser* 6, no. 2 (1924): 11–19.

———. "The Origin and Development of the Earth and Other Terrestrial Planets." *Geochimica et Cosmochimica Acta* 1, no. 4–6 (1951): 209–77.

———. "The Origin and Development of the Earth and Other Terrestrial Planets: A Correction." *Geochimica et Cosmochimica Acta* 2, no. 5–6 (1952): 263–68.

———. *The Planets: Their Origin and Development*. New Haven, CT: Yale University Press, 1952.

———. "Primary and Secondary Objects." *Journal of Geophysical Research* 64, no. 11 (1959): 1721–37.

———. "A Scientist Views the World Situation." *Bulletin of the Atomic Scientists*, February 15, 1946.

———. "Should America Build the H-Bomb?" *Bulletin of the Atomic Scientists* 6, no. 3 (March 1950): 72–73.

———. "Significance of the Hydrogen Isotopes." *Industrial & Engineering Chemistry* 26, no. 7 (July 1934): 803–6.

———. "Some Criticisms of 'On the Origin of the Lunar Surface Features' by G. P. Kuiper." *Proceedings of the National Academy of Sciences of the United States of America* 41, no. 7 (1955): 423.

———. "The Statistical Distribution of the Electrons in the Various Orbits of the Hydrogen Atom; Heat Capacities and Entropies of Diatomic and Polyatomic Gases." PhD diss., University of California, Berkeley, 1923.

———. "The Structure of the Hydrogen Molecule Ion." *Physical Review* 27, no. 2 (February 1, 1926): 216–37.

———. "The Teaching of Atomic Structure to Physical Chemists: Round-Table Discussion." *Journal of Chemical Education* 6, no. 2 (February 1929): 284–85.

———. "The Thermodynamic Properties of Isotopic Substances." *Journal of the Chemical Society*, 1947, 562–81.

Urey, Harold C., and Michael Amrine. "I'm a Frightened Man." *Collier's*, January 5, 1946.

Urey, Harold C., and F. Russell Bichowsky. "A Possible Explanation of the Relativity Doublets and Anomalous Zeeman Effects by Means of a Magnetic Electron." *Proceedings of the National Academy of Sciences of the United States of America* 12, no. 2 (February 1926): 80–85.

Urey, Harold C., and Charles A. Bradley. "Raman Spectra of Silico-Chloroform." *Physical Review* 37, no. 7 (April 1, 1931): 843.

Urey, Harold C., Ferdinand G. Brickwedde, and George M. Murphy. "A Hydrogen Isotope of Mass 2." *Physical Review* 39, no. 1 (January 1, 1932): 164–65.

———. "A Hydrogen Isotope of Mass 2 and Its Concentration." *Physical Review* 40, no. 1 (April 1, 1932): 1–15.

Urey, Harold C., and Helen Johnston. "The Absorption Spectrum of Chlorine Dioxide." *Physical Review* 38, no. 12 (December 15, 1931): 2131.

Urey, Harold C., Heinz A. Lowenstam, Samuel Epstein, and Charles R. McKinney. "Measurement of Paleotemperatures and Temperatures of the Upper Cretaceous of England, Denmark, and the Southeastern United States." *Bulletin of the Geological Society of America* 62 (April 1951): 399–416.

Urey, Harold C., and George M. Murphy. "The Relative Abundance of N14 and N15." *Physical Review* 38, no. 3 (August 1, 1931): 575.

US House of Representatives Committee on Un-American Activities. *Trial by Treason: The National Committee to Secure Justice for the Rosenbergs and Morton Sobell*. House Document, 85th Congress, 1st Session, 206. Washington, DC: US House of Representatives, 1956.

Van Nuys, Frank. *Americanizing the West: Race, Immigrants, and Citizenship, 1890–1930*. Lawrence: University Press of Kansas, 2002.

Waenke, Heinrich, and James R. Arnold. "Hans E. Suess. December 16, 1909–September 20, 1993." *Biographical Memoirs of the National Academies of Science* 87 (2006): 354–73.

Wang, Jessica. *American Science in an Age of Anxiety: Scientists, Anticommunism, and the Cold War*. Chapel Hill: University of North Carolina Press, 1999.

———. "Science, Security, and the Cold War: The Case of E. U. Condon." *Isis* 83, no. 2 (June 1992): 238–69.

Wang, Zuoyue. *In Sputnik's Shadow: The President's Science Advisory Committee and Cold War America*. New Brunswick, NJ: Rutgers University Press, 2008.

Warner, Deborah Jean. "Women Astronomers." *Natural History* 88, no. 5 (1979): 12–26.

Washburne, Robert. "Black Body Radiation and the Decomposition of Nitrogen Pentoxide in Molecular Beams." PhD diss., Johns Hopkins University, 1928.

Waterloo Press. "Thirteen Graduates." June 27, 1907.

Waters, Kay. "A Little Known Boyhood from the Past." *Glendoran*, June 1995.

Weart, William G. "Scientist Scores Secrecy on Atom." *New York Times*, October 22, 1949.

Westman, Robert S. "Two Cultures or One? A Second Look at Kuhn's The Copernican Revolution." *Isis* 85, no. 1 (1994): 79–115.

Westwick, Peter J. *The National Labs: Science in an American System, 1947–1974*. Cambridge, MA: Harvard University Press, 2003.

Whitaker, Ewen A. *The University of Arizona's Lunar and Planetary Labora-*

tory: Its Founding and Early Years. Tucson: University of Arizona Printing-Reproductions Department, 1985.

White, Frederick A. *Mass Spectrometry in Science and Technology*. New York: Wiley, 1968.

Whittemore, Gilbert. "World War I, Poison Gas Research, and the Ideals of American Chemists." *Social Studies of Science* 5, no. 2 (1975): 135.

Wilhelms, Don E. *To a Rocky Moon: A Geologist's History of Lunar Exploration*. Tucson: University of Arizona Press, 1993.

Wyckoff, Susan. "Anshen, Ruth Nanda (b. 1900)." Edited by Paula E. Hyman and Deborah D. Moore. *Jewish Women in America: An Historical Encyclopedia*. New York: Routledge, 1997.

Yost, Don M. Review of *Elements of Thermodynamics*, by D. Ter Haar and H. Wergeland. *Journal of the American Chemical Society* 89, no. 19 (September 1967): 5089.

Zuckerman, Harriet. "The Nobel Laureates in the United States: A Sociological Study of Scientific Collaboration." PhD diss., Columbia University, 1965.

———. *Scientific Elite: Nobel Laureates in the United States*. New York: Free Press, 1977.

Index

Page numbers in *italics* refer to figures.